BLACK+DECKER

The Complete Guide to

SHEDS

Updated 3rd Edition

Design & Build a Shed:
Complete Plans • Step-by-Step How-To

Cool
Springs

Quarto is the authority on a wide range of topics.

Quarto educates, entertains and enriches the lives of our readers—enthusiasts and lovers of hands-on living.

www.quartoknows.com

© 2017 Quarto Publishing Group USA Inc.
Text © 2017 Editors of Cool Springs Press
Photography © 2017 Richard Fleischman

First published in 2017 by Cool Springs Press, an imprint of Quarto Publishing Group USA Inc., 400 First Avenue North, Suite 400, Minneapolis, MN 55401 USA. Telephone: (612) 344-8100 Fax: (612) 344-8692

quartoknows.com
Visit our blogs at quartoknows.com

Cool Springs Press titles are also available at discounts in bulk quantity for industrial or sales-promotional use. For details contact the Special Sales Manager at Quarto Publishing Group USA Inc., 400 First Avenue North, Suite 400, Minneapolis, MN 55401 USA.

10 9 8 7 6 5 4 3 2 1

ISBN: 978-1-59186-673-2

Library of Congress Control Number: 2016959729

Acquiring Editor: Mark Johanson
Project Manager: Jordan Wiklund
Art Director: Brad Springer
Cover Designer: Brad Springer
Layout: Danielle Smith-Boldt
Photography: Rich Fleischman
Setbuilding: Brad Holden
Photo Assistance: Ian Miller

Cover photo courtesy of Outdoor Living Today
www.outdoorlivingtoday.com

Printed in China

The Complete Guide to Sheds
Created by: The Editors of Cool Springs Press, in cooperation with BLACK+DECKER.
BLACK & DECKER, BLACK+DECKER, the BLACK & DECKER and BLACK+DECKER logos and product names are trademarks of The Black & Decker Corporation, used under license. All rights reserved.

NOTICE TO READERS

For safety, use caution, care, and good judgment when following the procedures described in this book. The publisher and BLACK+DECKER cannot assume responsibility for any damage to property or injury to persons as a result of misuse of the information provided.

The techniques shown in this book are general techniques for various applications. In some instances, additional techniques not shown in this book may be required. Always follow manufacturers' instructions included with products, since deviating from the directions may void warranties. The projects in this book vary widely as to skill levels required: some may not be appropriate for all do-it-yourselfers, and some may require professional help.

Consult your local building department for information on building permits, codes, and other laws as they apply to your project.

The Complete Guide to
SHEDS

Updated 3rd Edition

Design & Build a Shed:
Complete Plans • Step-by-Step How-To

Cool
Springs
Press

Quarto is the authority on a wide range of topics.

Quarto educates, entertains and enriches the lives of our readers—enthusiasts and lovers of hands-on living.

www.quartoknows.com

© 2017 Quarto Publishing Group USA Inc.
Text © 2017 Editors of Cool Springs Press
Photography © 2017 Richard Fleischman

First published in 2017 by Cool Springs Press, an imprint of Quarto Publishing Group USA Inc., 400 First Avenue North, Suite 400, Minneapolis, MN 55401 USA. Telephone: (612) 344-8100
Fax: (612) 344-8692

quartoknows.com
Visit our blogs at quartoknows.com

10 9 8 7 6 5 4 3 2 1

ISBN: 978-1-59186-673-2

Library of Congress Control Number: 2016959729

Acquiring Editor: Mark Johanson
Project Manager: Jordan Wiklund
Art Director: Brad Springer
Cover Designer: Brad Springer
Layout: Danielle Smith-Boldt
Photography: Rich Fleischman
Setbuilding: Brad Holden
Photo Assistance: Ian Miller

Cover photo courtesy of Outdoor Living Today
www.outdoorlivingtoday.com

Printed in China

MIX
Paper from responsible sources
FSC® C101537

The Complete Guide to Sheds
Created by: The Editors of Cool Springs Press, in cooperation with BLACK+DECKER.
BLACK & DECKER, BLACK+DECKER, the BLACK & DECKER and BLACK+DECKER logos and product names are trademarks of The Black & Decker Corporation, used under license. All rights reserved.

NOTICE TO READERS

For safety, use caution, care, and good judgment when following the procedures described in this book. The publisher and BLACK+DECKER cannot assume responsibility for any damage to property or injury to persons as a result of misuse of the information provided.

The techniques shown in this book are general techniques for various applications. In some instances, additional techniques not shown in this book may be required. Always follow manufacturers' instructions included with products, since deviating from the directions may void warranties. The projects in this book vary widely as to skill levels required: some may not be appropriate for all do-it-yourselfers, and some may require professional help.

Consult your local building department for information on building permits, codes, and other laws as they apply to your project.

Contents

The Complete Guide to **Sheds**

Contents (Cont.)

Introduction

*B*efore you can pick the shed that is right for you, you need to consider a larger question: what is a shed?

There are actually lots of answers to that question. A shed can be a small, simple structure meant to hold a few hand tools near where you use them in the garden. Or it can be something a bit larger and more complex—a substitute garage where you park wheeled power tools and bulk yard supplies (or a real detached garage where you keep the family car). Of course, it can also be dressed up in fun colors, with interesting design details that make it the perfect backyard playhouse for kids—an incredibly interesting alternative to a tree fort. A handsome, well-appointed shed is the pretty package that holds all the mechanics and supplies for your pool without detracting from the stunning appearance of the water feature. Wired for power, a shed is a small (or not so small) guesthouse or home office. So the ultimate answer to the question is that a shed may be just about any structure you can imagine.

No matter what your particular answer is, building a shed that will perfectly serve your needs and do justice to your property and home means thoughtful planning and careful execution. Sheds live somewhere between a full-blown residential structure and a simple utility space. They are substantial enough to be ruled by codes in most cases, but small enough that anyone with even modest DIY skills can tackle one.

Choosing a shed begins with picking a design. Shed designs are unique combinations of the practical and the aesthetic. There must be enough space, configured correctly, to store what you need stored, or to serve the purpose you intend for the outbuilding. But any shed should also fit into its surroundings, and add to—rather than detract from—the look of the property. You can choose from a wide variety of prefab DIY shed designs. But you'll have the most control over a shed's look and floorplan by building and customizing it to your own tastes and needs. The projects section in the second half of this book presents both kits and custom designs. You'll find a wealth of options from the tiny to the huge.

Regardless of the shed you're building, success starts with the foundation. Most shed foundations are simpler than what you would need for a home, but you should always check with your local building department. Any shed larger than a closet most likely will be regulated by local codes dictating everything from where you can place the structure, to particulars about the construction. You won't have any problem meeting code requirements when you use the best practices and standards described in the pages that follow. We've included everything you'll need to know to assemble just about any outbuilding.

Along the way, you'll find useful information on outfitting your shed with features such as stairs or ramps, workbenches, windows, and more. Because, ultimately, no matter what constitutes a "shed," your shed should be exactly what you want it to be.

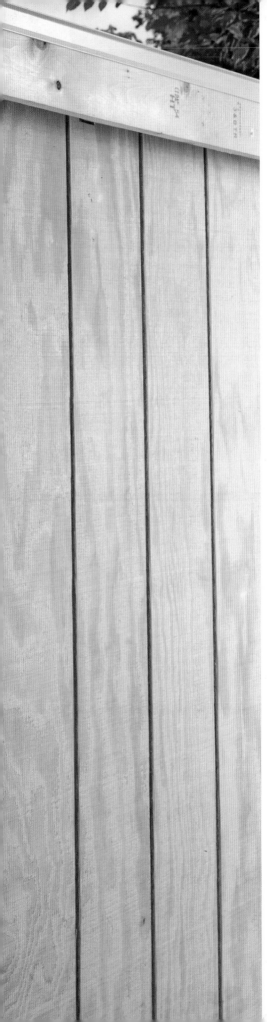

Building Basics

Choosing the size and style of your shed is just the first step in creating an outbuilding that will serve your purposes while doing credit to your existing landscape design. This section is all about making that vision a high-quality, code-compliant reality.

Although we begin with a discussion of shed options and choosing a site, most shed projects are best begun with a quick trip to the local building department. All but the smallest sheds, and any shed placed at or near your property line, will probably require a permit to build. Although they may seem like modest, safe structures, most sheds will need to conform to local codes. Failure to follow the codes may put you at risk of being forced to tear down an outbuilding you've put a lot of work into.

Once you're squared away on the municipal requirements, start from the bottom up. Turn to the pages on choosing and building a sound foundation for options that suit the size structure you've chosen and your local climate. You'll also find information on selecting the right materials for your shed, and on adding access features such as steps, decks and ramps, and more.

The key to a shed that looks great and is as useful as possible is getting these basics right. If you locate the shed on the best site and build it with the right materials on the right foundation, adding your own signature touches will be easy.

In this chapter:
- Choosing a Site for Your Shed
- Anatomy of a Shed
- Lumber & Hardware
- Building Foundations
- Ramps, Steps & Decks

Choosing a Site for Your Shed

The first step in choosing a site for your building doesn't take place in your backyard but at the local building and zoning departments. By visiting the departments, or making calls, you should determine a few things about your project before making any definite plans. Most importantly, find out whether your proposed building will be allowed by zoning regulations and what specific restrictions apply to your situation. Zoning laws govern such matters as the size and height of the building and the percentage of your property it occupies, the building's location, and its position relative to the house, neighboring properties, the street, etc.

From the building side of things, ask if you need a permit to build your structure. If so, you'll have to submit plan drawings (photocopied plans from this book should suffice), as well as specifications for the foundation and materials and estimated cost. Once your project is approved, you may need to buy a permit to display on the building site, and you may be required to show your work at scheduled inspections.

Because outbuildings are detached and freestanding, codes typically govern them loosely. Many impose restrictions or require permits only on structures larger than 100, or even 120, square feet. Others draw the line with the type of foundation used. In some areas, buildings with concrete slab or pier foundations are classified as "permanent" and thus are subject to a specific set of restrictions (and taxation, in some cases), while buildings that are set on skids and can—in theory at least—be moved are considered temporary or accessory and may be exempt from the general building codes.

Once you get the green light from the local authorities, you can tromp around your yard with a tape measure and stake your claim for the new building. Of course, you'll have plenty of personal and practical reasons for placing the building in a particular area, but here are a few general considerations to keep in mind:

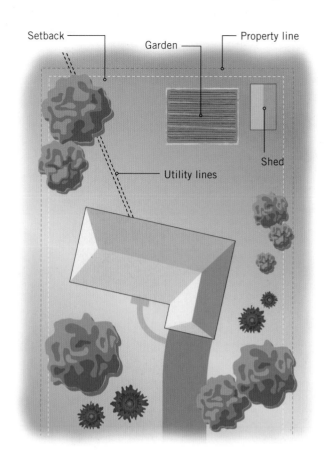

Visualize siting for your shed before it's built by simulating dimensions with stakes and a tarp. You can even move some of the items you plan to store into the staked-off area to see how they will fit.

Soil & drainage: To ensure that your foundation will last (whatever type it is), plant your building on solid soil, in an area that won't collect water.

Access: For trucks, wheelbarrows, kids, etc; do you want access in all seasons?

Utility lines: Contact local ordinances to find out where the water, gas, septic, and electrical lines run through your property. Often, local ordinances and utility companies require that lines are marked before digging. This is an essential step not only because of legalities, but also because you don't want your building sitting over lines that may need repair.

Setback requirements: Most zoning laws dictate that all buildings, fences, etc., in a yard must be set back a specific distance from the property line. This setback may range from 6 inches to 3 feet or more.

Neighbors: To prevent civil unrest, or even a few weeks of ignored greetings, talk to your neighbors about your project.

View from the house: Do you want to admire your handiwork from the dinner table, or would you prefer that your outbuilding blend in with the outdoors? A playhouse in plain view makes it easy to check on the kids.

Siting for Sunlight

Like houses, sheds can benefit enormously from natural light. Bringing sunlight into your backyard office, workshop, or storage structure makes the interior space brighter and warmer, and it's the best thing for combating a boxy feel. To make the most of natural light, the general rule is to orient the building so its long side (or the side with the most windows) faces south. However, be sure to consider the sun's position at all times of the year, as well as the shadows your shed might cast on surrounding areas, such as a garden or outdoor sitting area.

Seasonal Changes

Each day the sun crosses the sky at a slightly different angle, moving from its high point in summer to its low point in winter. Shadows change accordingly. In the summer, shadows follow the east-west axis and are very short at midday. Winter shadows point to the northeast and northwest and are relatively long at midday.

Generally, the south side of a building is exposed to sunlight throughout the year, while the north side may be shaded in fall, winter, and spring. Geographical location is also a factor: as you move north from the equator, the changes in the sun's path become more extreme.

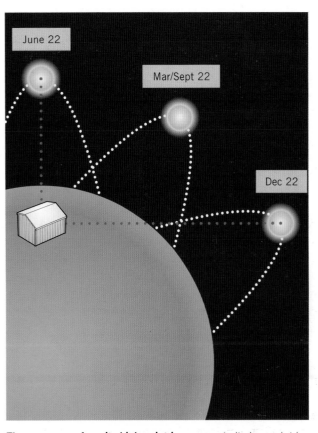

The sun moves from its high point in summer to its low point in winter. Shadows change accordingly.

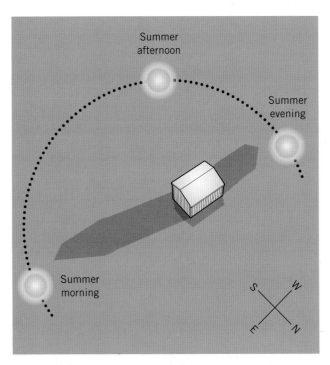

Shadows follow the east-west axis in the summer.

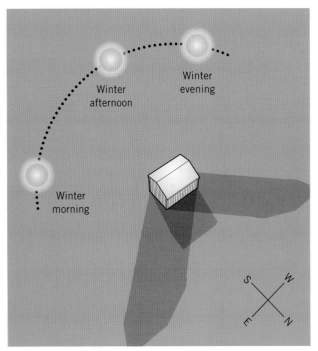

Winter shadows point to the northeast and northwest and are relatively long at midday.

Anatomy of a Shed

Shown as a cutaway, this shed illustrates many of the standard building components and how they fit together. It can also help you understand the major construction stages—each project in this book includes a specific construction sequence, but most follow the standard stages in some form:

1. Foundation—including preparing the site and adding a drainage bed;
2. Framing—the floor is first, followed by the walls, then the roof;
3. Roofing—adding sheathing, building paper, and roofing material;
4. Exterior finishes—including siding, trim, and doors and windows.

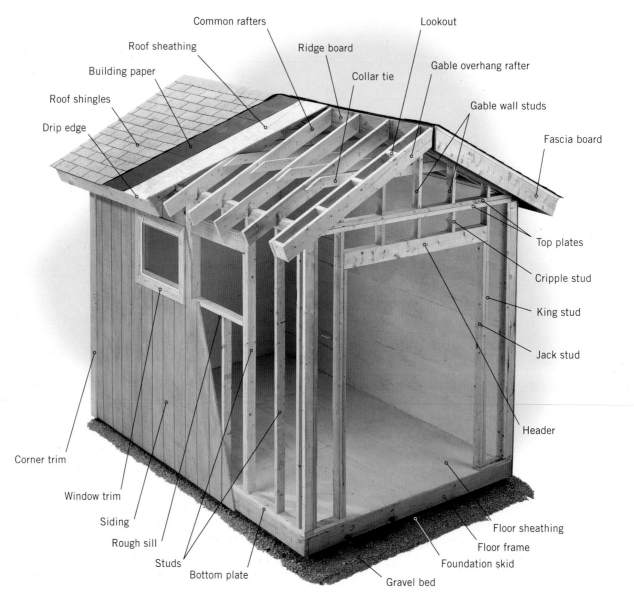

Common rafters
Roof sheathing
Building paper
Roof shingles
Drip edge
Ridge board
Collar tie
Lookout
Gable overhang rafter
Gable wall studs
Fascia board
Top plates
Cripple stud
King stud
Jack stud
Header
Floor sheathing
Floor frame
Foundation skid
Gravel bed
Bottom plate
Studs
Rough sill
Siding
Window trim
Corner trim

Lumber & Hardware

A combination of sheet stock, appearance-grade lumber, and structural lumber is used in most sheds.

The construction demands of most outbuildings are less rigorous than those of a more complex structure, such as a house. However, even a small shed should be built with sturdy structural elements to withstand time, insects, and the elements. That's why the lumber most commonly used in sheds is pressure-treated softwoods or a naturally resistant wood, such as cedar.

Pressure-treated lumber is your cheapest option, but also the least attractive. That makes it best for foundation and framing pieces that will be clad in another material. It's a good choice for sheds in which interior walls will be left unfinished and much of the wood will be left exposed to the elements and variations in temperature.

Modern pressure-treated wood contains highly corrosive chemicals. Always wear gloves when handling this lumber. And wear a dust mask and eye protection whenever you cut, drill, or sand pressure-treated lumber. Work on the wood outdoors, and never burn pressure-treated scraps.

Keep in mind that pressure-treated wood must be thoroughly dry before you paint or finish it. You can test how dry the wood is by sprinkling a few drops of water on the surface. The water should soak in immediately if the wood is dry enough for finishing. Even if you decide not to finish it, it's wise to treat the wood with clear wood preservative or waterproofing to prevent any potential water damage.

When choosing framing members, you'll pick a grade depending on where the member will be used. The grades—starting with the strongest and most attractive—range from *Select Structural* to *No. 1, No. 2,* and *No. 3. Stud grade* is used for load-bearing walls; and Standard and Construction grades are also suitable for this purpose, although less attractive. *Utility* and *Economy* grades are the least desirable, and should usually be avoided for use in sheds.

The finish lumber you'll use in visible locations is graded by quality and appearance, according to the number and size of knots. "Clear" means the number has no apparent knots. If you're painting your shed, a lesser grade may work fine for exterior surfaces.

All lumber has a *nominal dimension* (how it's labeled for sale) and an *actual dimension* (the true measurement). The chart on page 251 shows the differences for common lumber sizes. Any lumber greater than 4 inches thick is generally referred to as "timber." Lumber designated S4S (Surfaced-Four-Sides) has been milled smooth on all sides and follows standard dimensions. Boards with one or more rough faces can be over ⅛ inch thicker. Although you'll probably select exterior-grade plywood for any use in a shed, apply a finish to any exposed plywood edges to prevent water infiltration.

The right hardware is an essential complement to the wood you use in your shed. All the nails, screws, bolts, hinges, and anchors that will be exposed to the elements, rest on concrete, or come in contact with pressure-treated surfaces should be corrosion-resistant. The best are stainless steel or hot-dipped. Electroplated hardware is not as durable. More expensive stainless steel is the best choice for absolute durability and to avoid any hardware staining on cedar or redwood.

You may also need to use metal anchors or framing connectors to reinforce joints. If you have a hard time finding the right anchors or connectors at retail stores, check out manufacturers' websites for a more extensive selection of sizes and shapes. You can even order custom-made versions. Keep in mind that anchors and connectors are only effective if installed correctly. Follow the manufacturer's instructions to the letter, and use exactly the type and number recommended for your application.

Finally, finishing your shed will help prevent rot, insect damage, fading, and discoloration. Pine or other untreated lumber must have a protective finish if it will be exposed to the elements. But even cedar is susceptible to some rot and damage over time, and will turn a weathered gray if left unfinished. When painting wood, always prime it first to ensure the longest lasting coating.

Building Foundations

Your shed's foundation provides a level, stable structure to build upon and protects the building from moisture and erosion. In this section you'll learn to build four of the most common types of shed foundations. All but the concrete pier foundation are "on-grade" designs, meaning they are built on top of the ground and can be subject to rising and lowering a few inches during seasonal freezing and thawing of the underlying soil. This usually isn't a problem since a shed is a small, freestanding structure that's not attached to other buildings. However, it can adversely affect some interior finishes (wallboard, for example).

Frost-proof foundations are the better choice for larger, more finished sheds, or for those in colder parts of the country. The most common type of frost-proof shed foundation is poured concrete piers. Sheds are rarely, if ever, built on actual concrete foundation walls.

When choosing a foundation type for your shed, consider the specific site and the performance qualities of all systems in various climates; then check with the local building department to learn what's allowed and recommended in your area. Some foundations, such as concrete slabs, may classify sheds as permanent structures, which can affect property taxes, among other consequences. Residents in many areas may need to install special tie-downs or ground anchors according to local laws. If your building department requires a frost-proof foundation, you should be able to pass inspection by building your shed on concrete piers (see page 18).

Wooden Skid Foundation

A skid foundation couldn't be simpler: two or more treated wood beams or landscape timbers (typically 4 × 4, 4 × 6, or 6 × 6) set on a bed of gravel. The gravel provides a flat, stable surface that drains well to help keep the timbers dry. Once the skids are set, the floor frame is built on top of them, and nailed to the skids to keep everything in place.

Building a skid foundation is merely a matter of preparing the gravel base, then cutting, setting, and leveling the timbers. The timbers you use must be rated for ground contact. It is customary, but purely optional, to make angled cuts on the ends of the skids—these add a minor decorative touch and make it easier to skid the shed to a new location, if necessary.

Because a skid foundation sits on the ground, it is subject to slight shifting due to frost in cold-weather climates. Often a shed that has risen out of level will correct itself with the spring thaw, but if it doesn't, you can lift the shed with jacks on the low side and add gravel beneath the skids to level it.

TOOLS & MATERIALS

Shovel	Hand tamper	Treated wood timbers
Rake	Circular saw	Compactible gravel
4-ft. level	Square	Wood sealer-preservative
Straight, 8-ft. 2 × 4		

How to Build a Wooden Skid Foundation

Step 1: Prepare the Gravel Base

A. Remove 4" of soil in an area about 12" wider and longer than the dimensions of the building.

B. Fill the excavated area with a 4" layer of compactible gravel. Rake the gravel smooth, then check it for level using a 4-ft. level and a straight, 8-ft.-long 2 × 4. Rake the gravel until it is fairly level.

C. Tamp the gravel thoroughly using a hand tamper or a rented plate compactor. As you work, check the surface with the board and level, and add or remove gravel until the surface is level.

Step 2: Cut & Set the Skids

A. Cut the skids to length, using a circular saw or reciprocating saw. (Skids typically run parallel to the length of the building and are cut to the same dimension as the floor frame.)

B. To angle-cut the ends, measure down 1½" to 2" from the top edge of each skid. Use a square to mark a 45° cutting line down to the bottom edge, then make the cuts.

C. Coat the cut ends of the skids with a wood sealer-preservative and let them dry.

D. Set the skids on the gravel so they are parallel and their ends are even. Make sure the outer skids are spaced according to the width of the building.

Step 3: Level the Skids

A. Level one of the outside skids, adding or removing gravel from underneath. Set the level parallel and level the skid along its length, then set the level perpendicular and level the skid along its width.

B. Place the straight 2 × 4 and level across the first and second skids, then adjust the second skid until it's level with the first. Make sure the second skid is level along its width.

C. Level the remaining skids in the same fashion, then set the board and level across all of the skids to make sure they are level with one another.

Excavate the building site and add a 4" layer of compactible gravel. Level, then tamp the gravel with a hand tamper or rented plate compactor (inset).

2 × 4

If desired, mark and clip the bottom corners of the skid ends. Use a square to mark a 45° angle cut.

Using a board and a level, make sure each skid is level along its width and length, and is level with the other skids.

Concrete Block Foundation

Concrete-block shed foundations are easy and inexpensive to build. Solid concrete blocks are laid out in evenly spaced rows, stabilized and leveled, and the shed floor frame is then placed directly on top of the blocks. (Only solid concrete blocks are used for this type of foundation; blocks with cavities in them are prone to deterioration and crumbling over time, and under the stress of the shed's weight and exposure to the elements.) The blocks are placed on a gravel bed—either a large bed for the entire foundation, as shown here, or footprint beds under each block. The gravel keeps the foundation stable as the ground freezes and thaws, and drains water away from the foundation.

Concrete blocks are widely available in 4- and 2-inch thicknesses (thinner blocks are sold as "patio" blocks). The different thicknesses give you the necessary flexibility to create a level foundation over an irregular grade, such as a slope. Just stack the blocks in a configuration that creates an overall level surface.

MUD SILL BUFFER

A 2 × 8 mud sill adds strength to a standard 2 × 6 floor frame. First, you fasten the side rim joists to the sill, then you set the assembly on top of the foundation blocks and install the remaining floor joists.

TOOLS & MATERIALS

Mason's lines & stakes

Excavation tools

Hand tamper

2-ft. level

4-ft. level

Long, straight 2 × 4

Caulking gun

Compactible gravel

Solid concrete blocks

Asphalt shingles or 1 × 8 pressure-treated lumber, as needed

Construction adhesive

A foundation created with solid concrete blocks on a prepared base is simple to build and makes an easy solution to dealing with low slopes.

How to Build a Concrete Block Foundation

Outline the gravel bed footprint with lime or landscape paint. The bed should be at least 12" larger on all sides than the actual shed footprint. (The shed here is offset on the bed.) Remove 4" of soil and replace with compactable gravel. Compact with a hand tamper.

Measure and mark for the rows of blocks. Set and level the blocks, removing and adding gravel underneath each one as necessary. Use a level on a long 2 × 4 as a straightedge to ensure the blocks are level in the line.

When level, glue the stacked blocks together. Repeat with the next row, checking for level across the two rows. Glue the second line of blocks in place, once they have been leveled both ways.

Concrete Pier Foundation

Foundation piers are poured concrete cylinders that you form using cardboard tubes. The tubes come in several diameters and are commonly available from building materials suppliers. For an 8 × 10-foot shed, the minimal foundation consists of 8-inch-diameter piers at the corners.

You can anchor the shed's floor frame to the piers using a variety of methods. One method (shown here) is to bolt a wood block to the top of each pier, then fasten the floor frame to the blocks. Other anchoring options involve metal post bases and various framing connectors either set into the wet concrete or fastened to the piers after the concrete has cured.

The point of a frost-proof foundation such as this is to extend below the local frost line so that the foundation will not be subject to movement during freezing and thawing cycles. Any frost-proof foundation, including one formed from concrete piers, usually is regulated by local codes, which include specifics on how high the pier must project above the soil line, acceptable diameter of the pier in relation to the size of the shed, and how deep the piers must run to be truly "frost proof." In all cases, you should use a base of about 4 inches of gravel beneath each pier. But depending on local codes and conditions, you may need to reinforce the concrete with a rebar center post. Take care to keep the prefab tube plumb while pouring or adding the concrete.

How to Build a Concrete Pier Foundation

Circular saw	Plumb bob	2 × 4 lumber	Concrete mix
Drill	Shovel	2½" screws	J-bolts with washers
Mason's line	Posthole digger	Stakes	and nuts
Sledgehammer	Reciprocating saw	Nails	2 × 10 pressure-treated
Line level	Utility knife	Masking tape	lumber (rated for
Framing square	Ratchet wrench	Cardboard concrete forms	ground contact)

Cut batterboard pieces for 8 batterboards from 2 × 4s. Cut each leg to a point on one end. Use a square to ensure the legs are parallel and that the cross brace is perpendicular to the legs. Screw the batterboards together.

Tie mason lines to nails in the top center of the batterboards and run the lines as shown in the diagram below. Confirm that the lines are square using the 3-4-5 method (marking points with tape and level them with a line level).

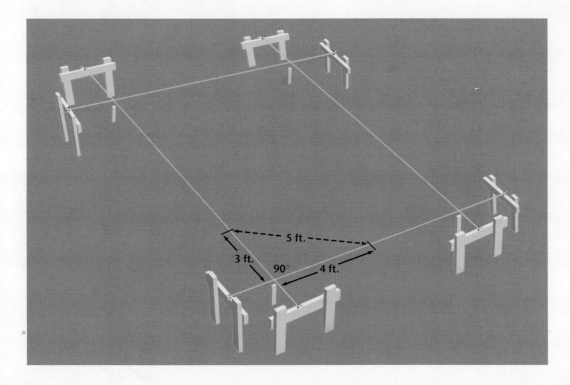

5 ft.

3 ft.

90°

4 ft.

(continued)

Use a plumb bob at the line intersections to mark each pier centerpoint. Drive a stake at that point. When all the piers are marked, untie one end of each line and coil the line and place it out of the way.

Dig holes centered on the stakes. Dig the holes a few inches larger in diameter than the pier forms and deep enough to comply with local codes. Add a base of gravel at least 4" thick. Cut the forms to extend at least 3" above ground, and make sure the tops of the piers are all level with one another. Set the tubes in the holes and fill around them with packed dirt.

Restring the lines and confirm that the placement of the forms is accurate. Pour the concrete, using a long stick to tamp and eliminate air pockets. Level the top of each so that it's flat, and set hardware, such as J-bolts, into the wet concrete. Use the plumb bob to ensure the hardware is centered.

Cut mounting blocks from pressure-treated lumber rated for ground contact. Drill a center hole, countersunk, in each block. (The hole must be large enough for the J-bolt to pass through.) Attach the blocks with galvanized nuts and washers, using the layout lines to align the blocks.

Concrete Slab Foundation

The slab foundation commonly used for sheds is called a slab-on-grade foundation. This combines a 3½- to 4-inch-thick floor slab with an 8- to 12-inch-thick perimeter footing that provides extra support for the walls of the building. The whole foundation can be poured at one time using a simple wood form.

Because they sit above ground, slab-on-grade foundations are susceptible to frost heave. Specific design requirements also vary by locality, so check with the local building department regarding the depth of the slab, the metal reinforcement required,

the type and amount of gravel required for the subbase, and whether plastic or another type of moisture barrier is needed under the slab.

The slab shown in this project has a 3½-inch-thick interior with an 8-inch-wide × 8-inch-deep footing perimeter. There is a 4-inch layer of compacted gravel underneath the slab and perimeter, and the concrete is reinforced internally with a layer of welded wire mesh (WWM). After the concrete is poured and finished, 8-inch galvanized J-bolts are set into the slab along the edges. These are used to anchor the shed wall framing to the slab.

 ## How to Build a Concrete Slab Foundation

TOOLS & MATERIALS

Circular saw	Concrete edger
Drill	Compactible gravel
Mason's line	2 × 3 & 2 × 4 lumber
Sledgehammer	1¼" & 2½" deck screws
Line level	¾" A-C plywood
Framing square	8d nails
Shovel	5 × 10-ft. welded wire
Wheelbarrow	mesh (WWM)
Rented plate compactor	1½" brick pavers
Bolt cutters	J-bolts
Bull float	2"-thick rigid
Hand-held finish float	foam insulation

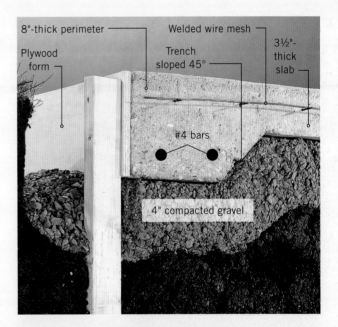

Labels: 8"-thick perimeter · Welded wire mesh · 3½"-thick slab · Plywood form · Trench sloped 45° · #4 bars · 4" compacted gravel

(1) Excavate the slab site and set up batterboards and lines as described on page 19. Measure down from the lines to check the depth of the excavation. It should be approximately 8" deep in the field, with a perimeter trench 8" wide × 12" deep (allowing for a 4" base of compacted gravel everywhere).

(2) Line the slab excavation with 4" of compacted gravel. Cut the form pieces from plywood or 1 × 8 lumber. Screw the pieces together and check to ensure that the inner dimensions of the form match the slab outer dimensions. For long runs, join the pieces with plywood mending plates. *(continued)*

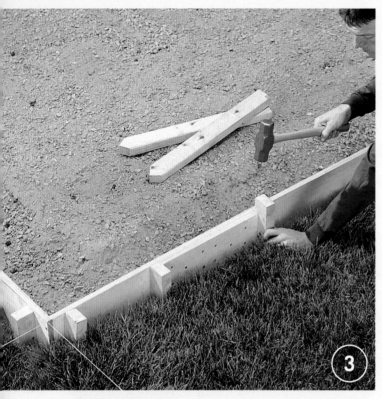

Set the form in place and stake it every 12", so that the form stands 4" above grade.

Lay out the wire mesh reinforcement in an even grid, propping up the mesh on 1½"-thick brick or stone bolsters. Mark the layout positions of the J-bolts on the edges of the form.

Estimate and order the concrete (see page 23). Pour the concrete into the form and then screed it flat and level, adding concrete to fill low spots. Rap the sides of the form to eliminate air pockets.

Float the slab with a bull float, and then set the J-bolts. Wait for the concrete to set up, and then float with a finish float and edge the slab. Allow the slab to cure for at least 24 hours before removing the forms, and then at least an additional 24 hours before building on the slab.

A slab for a shed requires a lot of concrete: an 8 × 10-ft. slab designed like the one in this project calls for about 1.3 cubic yards of concrete; a 12 × 12-ft. slab, about 2.3 cubic yards. Considering the amount involved, you'll probably want to order ready-mix concrete delivered by truck to the site (most companies have a minimum order charge). Tell the mixing company that you're using the concrete for an exterior slab.

An alternative for smaller slabs is to rent a concrete trailer from a rental center or landscaping company; they fill the trailer with one yard of mixed concrete and you tow it home with your own vehicle.

If you're having your concrete delivered, be sure to have a few helpers on-hand when the truck arrives; neither the concrete nor the driver will wait for you to get organized. Also, concrete trucks must be unloaded completely, so designate a dumping spot for any excess. Once the form is filled, load a couple of wheelbarrows with concrete (in case you need it) then have the driver dump the rest. Be sure to spread out and hose down the excess concrete so you aren't left with an immovable boulder in your yard.

If you've never worked with concrete, finishing a large slab can be a challenging introduction; you might want some experienced help with the pour.

Estimating Concrete

Calculate the amount of concrete needed for a slab of this design using this formula:

> Width × Length × Depth, in ft. (of main slab)
>
> Multiply by 1.5 (for footing edge and spillage)
>
> Divide by 27 (to convert to cubic yards)
>
> Example—for a 12 × 12-ft. slab:
>
> 12 × 12 × .29 (3½") = 41.76
>
> 41.76 × 1.5 = 62.64
>
> 62.64 ÷ 27 = 2.32 cubic yards

Timing is key to an attractive concrete finish. When concrete is poured, the heavy materials gradually sink, leaving a thin layer of water—known as bleed water—on the surface. To achieve an attractive finish, it's important to let bleed water dry before proceeding with other steps. Follow these rules to avoid problems:

- Settle and screed the concrete and add control joints immediately after pouring and before bleed water appears. Otherwise, crazing, spalling, and other flaws are likely.

- Let bleed water dry before floating or edging. Concrete should be hard enough that foot pressure leaves no more than a ¼"-deep impression.

- Do not overfloat the concrete; it may cause bleed water to reappear. Stop floating if a sheen appears, and resume when it is gone.

NOTE: Bleed water does not appear with air-entrained concrete, which is used in regions where temperatures often fall below freezing.

- Do not overload your wheelbarrow. Experiment with sand or dry mix to find a comfortable, controllable volume. This also helps you get a feel for how many wheelbarrow loads it will take to complete your project.

- Once concrete is poured and floated it must cure. It should not dry. If it is a hot day it is a good idea to spray mist from a hose after the concrete has "set" to keep it moist. Make sure you have a flat, stable surface between the concrete source and the forms.

- Start pouring concrete at the farthest point from the concrete source, and work your way back.

Ramps, Steps & Decks

Most of the sheds in this book with framed wood floors have a finished floor height at least 10 inches above the ground. This makes for a fairly tall step up to the shed. On a sloping site, the approach to the shed may be considerably lower than the floor. But not to worry—you can quickly build a custom ramp or set of steps for safe, easy access.

Add a single ramp to your shed to make moving heavy equipment in and out much easier.

Simple Ramp

A basic, sturdy ramp is a great convenience for moving heavy equipment in and out of your shed. Using the simple design shown here, you can make the slope of the ramp as gentle or as steep as you like (within reason). Of course, the gentler the slope, the easier it is roll things up the ramp. Construct your ramp from pressure-treated lumber rated for ground contact. If desired, set the bottom end of the ramp on a bed of compacted gravel for added stability and to ensure drainage away from the ramp.

 TOOLS & MATERIALS

Saw
Drill
Framing square
2 × 4 pressure-treated lumber
2 × 6 pressure-treated lumber
3" corrosion-resistant screws
Ruler
Pencil
Level

How to Build a Shed Ramp

Place a board between the ground and shed floor, experimenting to find the best angle. Measure up at the ground and make a mark 6" up the board, on the shed side.

①

②

Mark a level line 4⅝" below the floor, indicating the ledger's top edge. Cut the 2 × 4 ledger to the width of the ramp. Screw the ledger in place, centered under the door.

NOTE: The ramp should be at least as wide as the door opening.

Use a framing square to mark perpendicular lines—for the shed front height (minus 1⅝") and the ground, with a mark at the contact point—on a piece of plywood. Hold a 2 × 6 with the top edge touching both points. Mark for the angled ground end of the stringer, and the notch for the ledger. Make the cuts.

Dry-fit the first stringer and adjust as necessary. Use this first stringer as a template for the remaining stringers, using one stringer for every 12" of ramp width. Screw the stringers to the ledger and floor frame.

Cut 2 × 6 treads for the ramp. Screw the boards to the ledgers, leaving a ¼" gap between treads.

NOTE: Plan the tread layout so that the end treads are roughly the same width.

Traditional Stairs

A small set of framed wooden stairs is usually called for when a shed floor stands at about 21 inches or more above the ground (for lower floors, you might prefer to build a couple of simple platforms for three easy steps into the shed; see page 29). But regardless of the floor height, notched-stringer stairs add a nice handmade, built-in look to an entrance. And it's useful to learn the geometry and carpentry skills behind traditional stair building.

When planning your project, bear in mind that stairs in general are strictly governed by building codes. Your local building department may impose specific design requirements for your project, or they may not get involved at all—just be sure to find out. In any case, here are some of the standard requirements for stairs:

- Minimum tread depth: 10"

- Maximum riser height: 7¾" (7¼" is a good standard height)

- Minimum stair width: 36" (make your staircase at least a few inches wider than the shed's door opening)

- A handrail is often required for stairs with more than one riser, but this may not apply for storage sheds and the like (check with the local building department).

Because stairs are easier and safer to use when starting from a flat landing area, it's a good idea to include a level pad of compacted gravel at the base of your stairs. This also provides stability for the staircase and eliminates the potential for a slippery, muddy patch forming at the landing area.

CALCULATING STEP SIZE

Properly built stairs have perfectly uniform treads (the part you step on) and risers (the vertical section of each step).

To determine the riser height, all you have to do is divide the total rise—the distance from the ground to the shed floor—and divide by the number of steps. If you end up with risers over 7¾", add another step.

Determining the tread depth is up to you. However, because shallow steps are hard to climb and easy to trip on, you should make your treads at least 10" deep, but preferably 11" or more. The tread depth multiplied by the number of steps gives you the total run—how far the steps extend in front of the shed. When cutting the stringers, you cut the first (bottom) riser shorter than the others to account for the thickness of the tread material.

Stringer stairs are an easy and often necessary addition to elevated shed doors.

How to Build Notched-Stringer Stairs

Step 1: Lay Out the First Stringer

A. Use a framing square to lay out the first stringer onto a straight piece of 2 × 12 lumber. Starting at one end of the board, position the square along the board's top edge. Align the 12" mark of the blade (long leg of the square) and the 6½" mark on the tongue (short leg of the square) with the edge of the board. Trace along the edges of the blade and tongue. The tongue mark represents the first riser.

B. Use the square to extend the blade marking across the full width of the board. Then, draw a parallel line 1" up from this line. The new line marks the bottom cut for the stringer (the 1" offset accounts for the thickness of the tread material).

C. Continue the step layout, starting at the point where the first riser meets the top of the board.

D. Mark the cutting line at the top end of the stringer by extending the third (top) tread marking across the full width of the board. From this line, make a perpendicular line 12" from the top riser: this is where you'll cut the edge of the stringer that fits against the shed.

Step 2: Cut the Stringers

A. Make the cuts on the first stringer using a circular saw set to full depth. Where treads and risers intersect, cut just up to the lines, then finish the cuts with a handsaw.

B. Test-fit the stringer on the shed. The top tread cut should be 1" below the shed floor. Make adjustments to the stringer cuts as needed.

C. Use the first stringer as a template to mark the remaining stringers. You'll need one stringer for each end and every 12" to 16" in between. Cut the remaining stringers.

TOOLS & MATERIALS

Framing square	10d × 1½" galvanized nails
Circular saw	3" galvanized screws
Handsaw	2½" galvanized screws
Drill and bits	2 × 4 pressure-treated lumber
Sledgehammer	
Compactible gravel	2 16" lengths of #4 (½"-dia.) rebar
2 × 12 pressure-treated lumber	5/4 × 6 pressure-treated lumber
Corrosion resistant framing connectors	

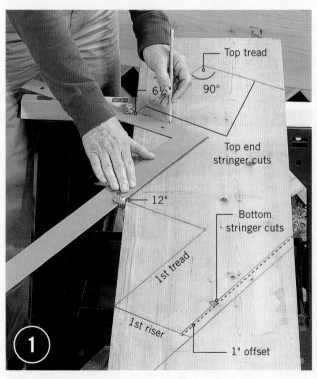

Top tread • 90°

6½"

Top end stringer cuts

12"

Bottom stringer cuts

1st tread

1st riser

1" offset

Use a framing square to mark the treads, risers, and end cuts on the stringer.

Cut out stringers with a circular saw, and finish the corners of the cuts with a handsaw. *(continued)*

Step 3: Install the Stringers

A. Mark a level line onto the shed's floor frame, 1" below the finished floor surface. Onto the level line, mark the center and outsides of the stairs. Transfer the side markings to the ground using a square and straightedge.

B. Cut 2× blocking to carry the bottom ends of the stringers. Fasten the blocking to the ground using 16" pieces of #4 rebar driven through ½" holes. For concrete or masonry, use masonry screws or a powder-actuated nailer.

C. Anchor the tops of the stringers to the floor frame with corrosion-resistant framing connectors, using 1½"-long 10d galvanized common nails. The tops of the stringers should be flush with the level line.

D. Anchor the bottom ends of the stringers with nails or screws.

Step 4: Add the Tread & Riser Boards

A. Cut the treads to length from 2 × 8" pressure-treated decking lumber or two 2 × 6" boards ripped to 4" each. You can cut the treads to fit flush with the outside stringers or overhang them by ½" or so for a different look.

B. On each step, position a full-width tread at the front with the desired overhang beyond the riser below. Fasten the tread to the stringers with 2½" galvanized screws. Rip the second tread to size and fasten it behind the first tread, leaving a ¼" gap between the boards. Install the remaining tread boards.

C. If desired, install 1 × 6 riser boards so the ends are flush with the outside stringers (no overhang). Or, you can omit the riser boards for open steps.

Mount the top ends of the stringers to the shed using framing connectors (shown). Anchor the bottom ends to blocking.

Trim the rear tread boards as needed to fit behind the front treads.

Risers are not necessary to maintain the structural integrity of the stairs. Stairs that don't include risers are called "open staircases," while those with risers are called "enclosed." You can save a modest amount of material costs and fabrication time by leaving your staircase open, but enclosed stairs look much more polished and finished.

Platforms for Steps & Decking

This simple platform is a popular option for sheds because it's so easy to build and it provides a sturdy step for comfortable access. You can use the same basic design to make platforms of any size. A large platform can become an outdoor sitting area, while a stack of smaller platforms can create a set of steps that are accessible from three directions. For stability and longevity, set your platforms on top of solid concrete blocks—the same type used for block foundations. See page 16 for more information about building with concrete block.

TOOLS & MATERIALS

Shovel	Compactible gravel (optional)
Level	
Saw	16d galvanized common nails
Drill	
2× treated lumber	3" deck screws

How to Build a Basic Platform

Step 1: Build the Platform Frame

A. Cut two long side pieces from 2 × 6 lumber. These should equal the total length of the frame. Cut two end pieces to fit between the side pieces. For example, if your platform will measure 24 × 36", cut the sides at 36" and cut the ends at 23". Also cut an intermediate support for every 16" in between: make these the same length as the end pieces.

B. Fasten the end pieces between the sides with pairs of 16d galvanized common nails.

C. Fasten the intermediate support at the center of the frame or at 16" intervals.

Step 2: Install the Decking

A. Cut 2 × 6 decking boards to fit the long dimension of the platform frame (you can also use 5/4 × 6 decking boards).

OPTION: For a finished look, cut the decking about 1" too long so it overhangs the frame structure.

B. Measure the frame diagonally from corner to corner to make sure it is square.

C. Starting at the front edge of the frame, attach the decking to the framing pieces with pairs of 3" deck screws. Leave a ¼" gap between the boards. Rip the last board to width so that it overhangs the front edge of the frame by 1".

D. Set the platform in position on top of the block foundation. If desired, fasten the platform to the shed with 3" screws.

Assemble the frame pieces with pairs of 16d common nails.

Install the decking with screws, leaving a ¼" gap between boards.

Shed Projects

It's time to get that tool belt dusty—it's time to build a shed. Each custom project in this section features a complete materials list to make shopping at your local home center and lumberyard a breeze. From the detailed drawings and how-to instructions you'll learn what length to cut each piece and where it goes in the finished product. The step-by-step instructions will walk you through the entire sequence, highlighting important and unique details along the way. Even if you don't find the exact shed you want in this chapter, you can easily alter the plans or even combine elements from different sheds to one that meets your needs perfectly.

NOTE: Making sheds larger than shown may require review of code/design criteria for lumber sizes, especially floor joists, roof joists, and headers.

In this chapter:

Small Sheds
- Lean-To Tool Shed
- Mini Garden Shed
- Garbage & Recycling Shed
- Call Box Shed
- Bike Shed

Medium Sheds
- Modern Utility Shed
- Simple Storage Shed
- Timber-Frame Shed
- Gothic Playhouse
- Salt Box Storage Shed

Large Sheds
- Clerestory Studio
- Sunlight Garden Shed
- Gambrel Garage
- Convenience Shed

Kit Sheds
- Building a Metal or Wood Kit Shed

Small Sheds

Smaller sheds are quickly growing in popularity. They're attractive to homeowners because they're easy to build and less expensive than a larger structure. Whether you buy a kit or build custom, the project is likely to take you less than a weekend from start to finish. But just because you opt for something more modest doesn't mean you have to give up an attractive design. The real choice with small sheds (sometimes called "mini" or "micro" sheds) is whether to blend them in with their surroundings or make them a focal point of a more sedate outdoor area. Either way, look for ways to customize the shed so that it does exactly what you want within its small footprint.

Focus on function in smaller sheds. Sometimes, you only need a shed for one role, such as concealing unattractive garbage and recycling bins. This tiny cedar unit serves that basic function admirably, with openings on the front and top that make accessing the bins easy. But it also creates a handsome visual that fits right in next to a swimming pool.

Choose understatement where the shed needs to seamlessly blend with its surroundings. This uncomplicated, cedar prefab shed combines a pleasing monocolor design with a minimal architectural style to create a storage space that doesn't overwhelm anything else on the patio.

Optimize access with barn doors. This homey arched shed brightens the garden with its sun-yellow exterior. But function trumps form with the full-front barn doors that allow complete and easy access to everything stored in the shed.

Integrate a small shed into its surroundings with features that blend. It's easy to overlook the finish on your shed, but that's a mistake. The paint color used on this distinctive small shed matches the wood arbor and slate patio. The look is pleasing and calm.

Max out micro spaces with a shed sized to fit. A tiny shed like this one can provide enclosed storage for essentials without getting in the way of foot traffic. Smaller sheds are generally inexpensive, incredibly easy to build, and the best solution for a temporary situation such as a rental, where you'll probably need to move the shed in the future.

Configure a small shed to meet your needs. Even a small space needs to be organized for you to get the most out of it. This shed includes shelves for smaller items while still leaving plenty of space for long-handled tools. Hang hooks on the back of the door for even more storage.

Go modern for big bang in a small shed. This backyard stunner is an example of using big style in a small structure, and it adds a showpiece to an otherwise mundane brick house wall and fieldstone patio. The double doors add plenty of access to the storage space inside.

Simple is best for small sheds. A basic design without much ornamentation ensures a small shed fits wherever you need it to go. The uncomplicated design also requires a minimum of material costs and effort to build.

Double down for finest function. Double doors are as useful on a small shed as they are on a larger structure. This simple wood shed provides ample access with full-swing double doors and a solid path to the doors for wheeled yard equipment.

Lean-To Tool Shed

The lean-to is a classic outbuilding intended as a supplementary structure for a larger building. Its simple shed-style roof helps it blend with the neighboring structure and directs water away and keeps leaves and debris from getting trapped between the two buildings. When built to a small shed scale, the lean-to (sometimes called a closet shed) is most useful as an easy-access storage locker that saves you extra trips into the garage for often-used lawn and garden tools and supplies.

This lean-to tool bin is not actually attached to the house, though it appears to be. It is designed as a freestanding building with a wooden skid foundation that makes it easy to move. With all four sides finished, the bin can be placed anywhere, but it works best when set next to a house or garage wall or a tall fence. If you locate the bin out in the open—where it won't be protected against wind and extreme weather—be sure to anchor it securely to the ground to prevent it from blowing over.

As shown here, the bin is finished with asphalt shingle roofing, T1-11 plywood siding, and 1× cedar trim, but you can substitute any type of finish to match or complement a neighboring structure. Its 65-inch-tall double doors provide easy access to its 18 square feet of floor space. The 8-foot-tall rear wall can accommodate a set of shelves while leaving enough room below for long-handled tools.

Because the tool bin sits on the ground, in cold climates it will be subject to shifting with seasonal freeze-thaw cycles. Therefore, do not attach the tool bin to your house or any other building set on a frost-proof foundation.

Keep your tools safe and dry in the lean-to tool bin located next to a house, garage, fence or wall.

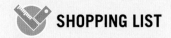

SHOPPING LIST

DESCRIPTION	QTY/SIZE	MATERIAL
FOUNDATION		
Drainage material	0.5 cu. yd.	Compactible gravel
Skids	2 @ 6'	4 × 4 treated timbers
FLOOR FRAMING		
Rim joists	2 @ 6'	2 × 6 pressure treated
Joists	3 @ 8'	2 × 6 pressure treated
Floor sheathing	1 sheet @ 4 × 8	¾" ext.-grade plywood
Joist clip angles	4	3 × 3 × 3" × 16-gauge galvanized
WALL FRAMING		
Bottom plates	1 @ 8', 2 @ 6'	2 × 4
Top plates	1 @ 8', 3 @ 6'	2 × 4
Studs	14 @ 8', 8 @ 6'	2 × 4
Header	2 @ 6'	2 × 6
Header spacer	1 piece @ 6'	½" plywood—5" wide
ROOF FRAMING		
Rafters	6 @ 6'	2 × 6
Ledger*	1 @ 6'	2 × 6
ROOFING		
Roof sheathing	2 sheets @ 4 × 8'	½" ext.-grade plywood
Shingles	30 sq. ft.	250# per square min.
Roofing starter strip	7 linear ft.	
15# building paper	30 sq. ft.	
Metal drip edge	24 linear ft.	Galvanized metal
Roofing cement	1 tube	
EXTERIOR FINISHES		
Plywood siding	4 sheets @ 4 × 8'	⅝" Texture 1-11 plywood siding, grooves 8" O.C.
Door trim	2 @ 8'	1 × 10 S4S cedar
	2 @ 6'	1 × 8 S4S cedar
Corner trim	6 @ 8'	1 × 4 S4S cedar
Fascia	3 @ 6'	1 × 8 S4S cedar
	1 @ 6'	1 × 4 S4S cedar
Bug screen	8" × 6'	Fiberglass

DESCRIPTION	QTY/SIZE	MATERIAL
DOORS		
Frame	3 @ 6'	¾" × 3½" (actual) cedar
Stops	3 @ 6'	1 × 2 S4S cedar
Panel material	12 @ 6'	1 × 6 T&G V-joint S4S cedar
Z-braces	2 @ 10'	1 × 6 S4S cedar
Construction adhesive	1 tube	
Interior trim (optional)	3 @ 6'	1 × 3 S4S cedar
Strap hinges	6, with screws	
FASTENERS		
16d galvanized common nails	3½ lbs.	
16d common nails	3½ lbs.	
10d common nails	12 nails	
10d galvanized casing nails	20 nails	
8d galvanized box nails	½ lb.	
8d galvanized finish nails	2 lbs.	
8d common nails	24 nails	
8d box nails	½ lb.	
1½" joist hanger nails	16 nails	
⅞" galvanized roofing nails	¼ lb.	
2½" deck screws	6 screws	
1¼" wood screws	60 screws	

---

*NOTE: 6-foot material is often unavailable at local lumber stores, so buy half as much of 12-foot material.

FLOOR FRAMING PLAN

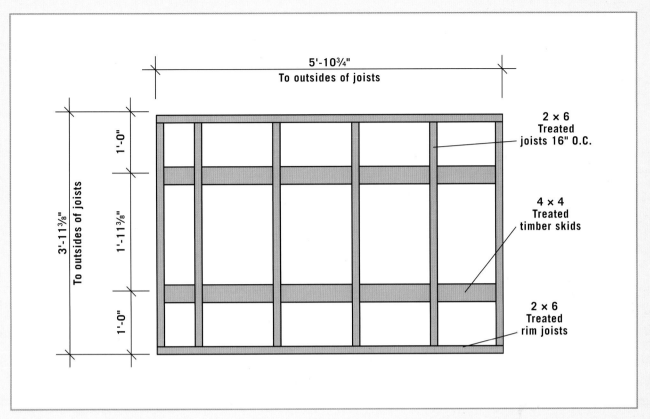

5'-10¾"
To outsides of joists

3'-11⅜"
To outsides of joists

1'-0"

1'-11⅜"

1'-0"

2 × 6
Treated
joists 16" O.C.

4 × 4
Treated
timber skids

2 × 6
Treated
rim joists

ROOF FRAMING PLAN

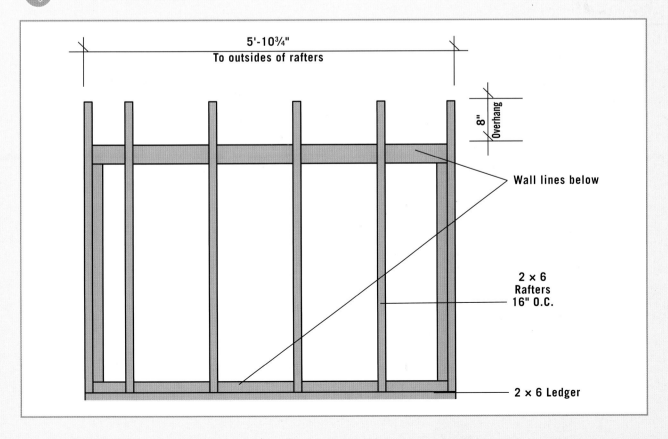

5'-10¾"
To outsides of rafters

8" Overhang

Wall lines below

2 × 6
Rafters
16" O.C.

2 × 6 Ledger

FRONT FRAMING ELEVATION

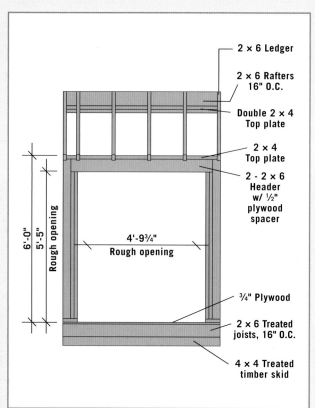

2 × 6 Ledger

2 × 6 Rafters
16" O.C.

Double 2 × 4
Top plate

2 × 4
Top plate

2 - 2 × 6
Header
w/ ½"
plywood
spacer

¾" Plywood

2 × 6 Treated
joists, 16" O.C.

4 × 4 Treated
timber skid

6'-0"

5'-5"
Rough opening

4'-9¾"
Rough opening

LEFT FRAMING ELEVATION

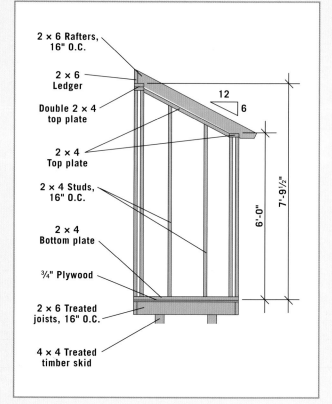

2 × 6 Rafters,
16" O.C.

2 × 6
Ledger

Double 2 × 4
top plate

2 × 4
Top plate

2 × 4 Studs,
16" O.C.

2 × 4
Bottom plate

¾" Plywood

2 × 6 Treated
joists, 16" O.C.

4 × 4 Treated
timber skid

12

6

7'-9½"

6'-0"

REAR FRAMING ELEVATION

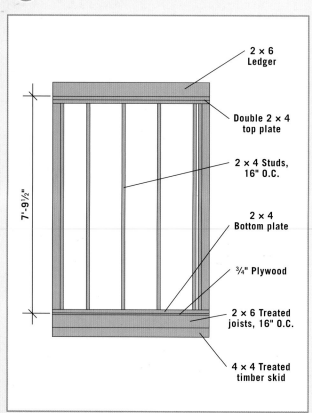

2 × 6
Ledger

Double 2 × 4
top plate

2 × 4 Studs,
16" O.C.

2 × 4
Bottom plate

¾" Plywood

2 × 6 Treated
joists, 16" O.C.

4 × 4 Treated
timber skid

7'-9½"

RIGHT FRAMING ELEVATION

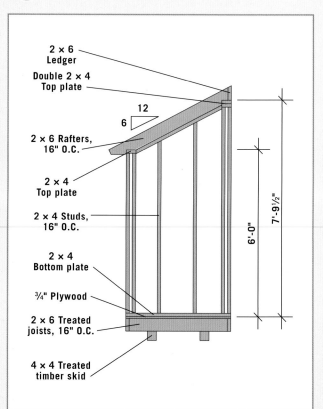

2 × 6
Ledger

Double 2 × 4
Top plate

2 × 6 Rafters,
16" O.C.

2 × 4
Top plate

2 × 4 Studs,
16" O.C.

2 × 4
Bottom plate

¾" Plywood

2 × 6 Treated
joists, 16" O.C.

4 × 4 Treated
timber skid

12

6

7'-9½"

6'-0"

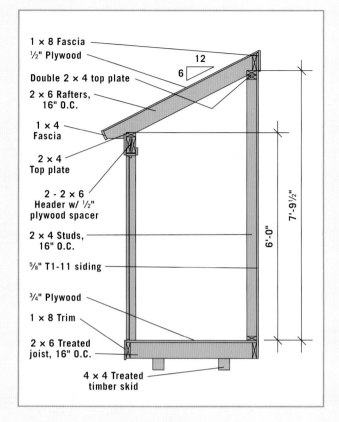

1 × 8 Fascia

½" Plywood

12

6

Double 2 × 4 top plate

2 × 6 Rafters, 16" O.C.

1 × 4 Fascia

2 × 4 Top plate

2 - 2 × 6 Header w/ ½" plywood spacer

2 × 4 Studs, 16" O.C.

⅝" T1-11 siding

¾" Plywood

1 × 8 Trim

2 × 6 Treated joist, 16" O.C.

4 × 4 Treated timber skid

7'-9½"

6'-0"

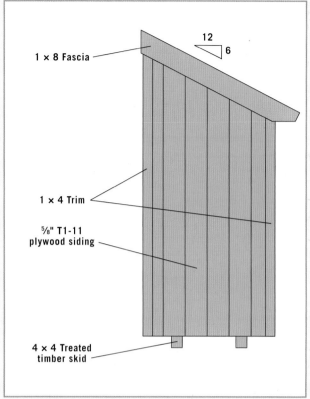

1 × 8 Fascia

12

6

1 × 4 Trim

⅝" T1-11 plywood siding

4 × 4 Treated timber skid

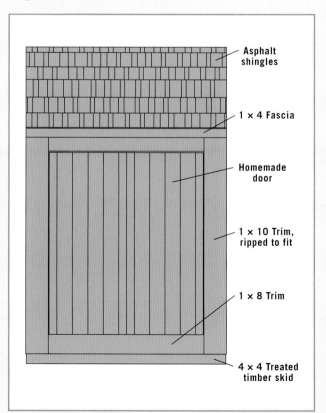

Asphalt shingles

1 × 4 Fascia

Homemade door

1 × 10 Trim, ripped to fit

1 × 8 Trim

4 × 4 Treated timber skid

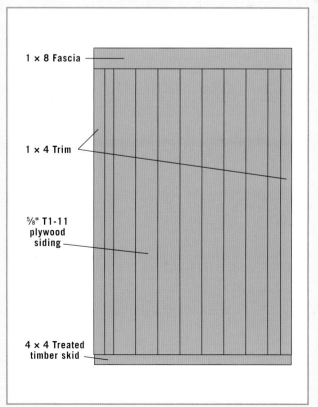

1 × 8 Fascia

1 × 4 Trim

⅝" T1-11 plywood siding

4 × 4 Treated timber skid

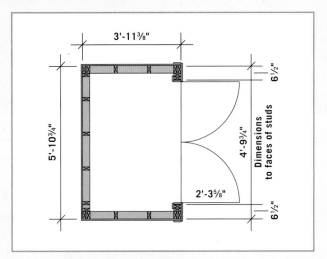

3'-11⅜"

5'-10¾"

4'-9¾"

Dimensions to faces of studs

6½"

6½"

2'-3⅝"

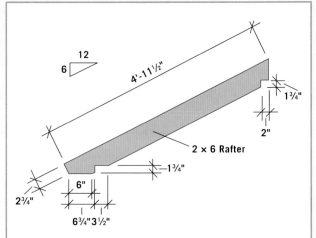

12

6

4'-11½"

1¾"

2"

2 × 6 Rafter

1¾"

2¾"

6"

6¾" 3½"

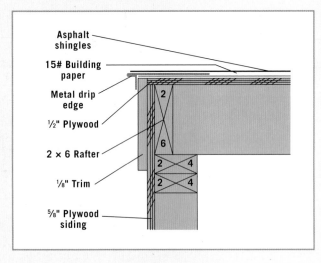

Asphalt shingles

15# Building paper

Metal drip edge

½" Plywood

2 × 6 Rafter

⅛" Trim

⅝" Plywood siding

2
6
2 4
2 4

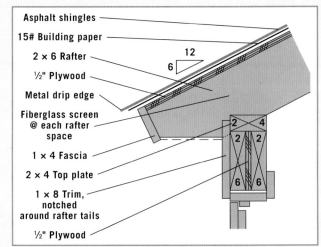

Asphalt shingles

15# Building paper

2 × 6 Rafter

½" Plywood

Metal drip edge

Fiberglass screen @ each rafter space

1 × 4 Fascia

2 × 4 Top plate

1 × 8 Trim, notched around rafter tails

½" Plywood

12
6

2 4
2 2
2
6 6

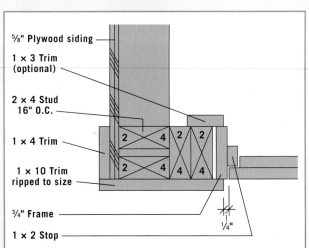

⅝" Plywood siding

1 × 3 Trim (optional)

2 × 4 Stud 16" O.C.

1 × 4 Trim

1 × 10 Trim ripped to size

¾" Frame

1 × 2 Stop

2 4 2 2
2 4 4 4

¼"

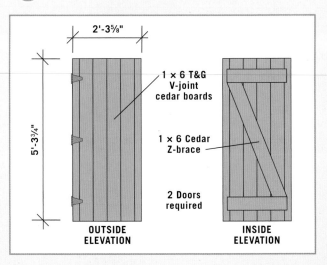

2'-3⅝"

5'-3¾"

1 × 6 T&G V-joint cedar boards

1 × 6 Cedar Z-brace

2 Doors required

OUTSIDE ELEVATION

INSIDE ELEVATION

How to Build the Lean-to Tool Shed

Prepare the site with a 4" layer of compacted gravel. Cut the two 4 × 4 skids at 70¾". Set and level the skids following FLOOR FRAMING PLAN (page 39). Cut two 2 × 6 rim joists at 70¾" and six joists at 44⅜". Assemble the floor and set it on the skids as shown in the FLOOR FRAMING PLAN. Check for square, and then anchor the frame to the skids with four joist clip angles (inset photo). Sheath the floor frame with ¾" plywood.

Cut plates and studs for the walls: Side walls—two bottom plates at 47⅜", four studs at 89", and four studs at 69"; Front wall—one bottom plate at 63¾", one top plate at 70¾", and four jack studs at 63½". Rear wall—one bottom plate at 63¾", two top plates at 70¾", and six studs at 89". Mark the stud layouts onto the plates.

Fasten the four end studs of each side wall to the bottom plate. Install these assemblies. Construct the built-up 2 × 6 door header at 63¾". Frame and install the front and rear walls, leaving the top plates off at this time. Nail together the corner studs, making sure they are plumb. Install the rear top plates flush to the outsides of the sidewall studs. Install the front top plate in the same fashion.

Cut the six 2 × 6 rafters following the RAFTER TEMPLATE (page 42). Cut the 2 × 6 ledger at 70¾" and bevel the top edge at 26.5° so the overall width is 4⁵⁄₁₆". Mark the rafter layout onto the wall plates and ledger, as shown in the ROOF FRAMING PLAN (page 39), then install the ledger flush with the back side of the rear wall. Install the rafters.

(continued)

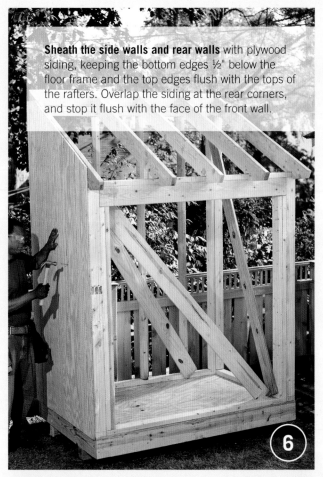

Complete the sidewall framing: Cut a top plate for each side to fit between the front and rear walls, mitering the ends at 26.5°. Install the plates flush with the outsides of the end rafters. Mark the stud layouts onto the sidewall top plates, then use a plumb bob to transfer the marks to the bottom plate. Cut the two studs in each wall to fit, mitering the top ends at 26.5°. Toenail the studs.

Sheath the side walls and rear walls with plywood siding, keeping the bottom edges ½" below the floor frame and the top edges flush with the tops of the rafters. Overlap the siding at the rear corners, and stop it flush with the face of the front wall.

Add the 1 × 4 fascia over the bottom rafter ends as shown in the OVERHANG DETAIL (page 42). Install 1 × 8 fascia over the top rafter ends. Overhang the front and rear fascia to cover the ends of the side fascia, or plan to miter all fascia joints. Cut the 1 × 8 side fascia to length, cut the front bottom corner level with the bottom of the front fascia. Install the side fascia.

Install the ½" roof sheathing, starting with a full-width sheet at the bottom edge of the roof. Fasten metal drip edge along the front edge of the roof. Cover the roof with building paper, then add the drip edge along the sides and top of the roof. Shingle the roof, and finish the top edge with cut shingles or a solid starter strip.

Cut and remove the bottom plate inside the door opening. Cut the 1 × 4 head jamb for the door frame at 57⅝" and cut the side jambs at 64". Fasten the head jamb over the sides with 2½" deck screws. Install 1 × 2 door stops ¾" from the front edges of jambs, as shown in the DOOR JAMB DETAIL (page 42). Install the frame in the door opening, using shims and 10d casing nails.

For each door, cut six 1 × 6 tongue-and-groove boards at 63¾". Fit them together, then mark and trim the two end boards so the total width is 27⅝". Cut the 1 × 6 Z-brace boards following the DOOR ELEVATION DETAIL (page 42). The ends of the horizontal braces should be 1" from the door edges. Attach the braces with construction adhesive and 1¼" screws. Install each door with three hinges.

Staple fiberglass insect mesh along the underside of the roof from each side 2 × 6 rafter. Cut and install the 1 × 8 trim above the door, overlapping the side door jambs about ¼" on each side (see the OVERHANG DETAIL, page 42).

Rip vertical and horizontal trim boards to width, then notch them to fit around the rafters, as shown in the DOOR JAMB DETAIL (page 42). Notch the top ends of the 1 × 10s to fit between the rafters and install them. Add 1 × 8 trim horizontally between the 1 × 10s below the door. Install the 1 × 4 corner trim, overlapping the pieces at the rear corners.

Mini Garden Shed

Whether you are working in a garden or on a construction site, getting the job done is always more efficient when your tools are close at hand. Offering just the right amount of on-demand storage, this mini garden shed can handle all of your gardening hand tools but with a footprint that keeps costs and labor low.

The mini shed base is built on two 2 × 8 front and rear rails that raise the shed off the ground. The rails can also act as skids, making it possible to drag the shed like a sled after it is built. The exterior is clad with vertical-board-style fiber-cement siding. This type of siding not only stands up well to the weather, but it is also very stable and resists rotting and warping. It also comes preprimed and ready for painting. Fiber-cement siding is not intended to be in constant contact with moisture, so the manufacturer recommends installing it at least 6 inches above the ground. You can paint the trim and siding any color you like. You might choose to coordinate the colors to match your house, or you might prefer a unique color scheme so that the shed stands out as a garden feature.

The roof is made with corrugated fiberglass roof panels. These panels are easy to install and are available in a variety of colors, including clear, which will let more light into the shed. An alternative to the panels is to attach plywood sheathing and then attach any roofing material you like over the sheathing. These plans show how to build the basic shed, but you can customize the interior with hanging hooks and shelves to suit your needs.

WORKING WITH FIBER-CEMENT SIDING

Fiber-cement siding is sold in different profiles and sheets at many home centers. There are specially designed shearing tools that contractors use to cut this material, but you can also cut it by scoring it with a utility knife and snapping it—just like cement tile backer board or drywall board. You can also cut cementboard with a circular saw, but you must take special precautions. Cementboard contains silica. Silica dust is a respiratory hazard. If you choose to cut it with a power saw, then minimize your dust exposure by using a manufacturer-designated saw blade designed to create less fine dust and by wearing a NIOSH/MSHA-approved respirator with a rating of N95 or better.

This scaled-down garden shed is just small enough to be transportable. Locate it near gardens or remote areas of your yard where on-demand tool storage is useful.

SHOPPING LIST

KEY	PART	DIMENSION	PCS.	MAT.
LUMBER				
A	Front/rear base rails	1½ × 7¼ × 55"	2	Treated pine
B	Base crosspieces	1½ × 3½ × 27"	4	Treated pine
C	Base platform	¾ × 30 × 55"	1	Ext. plywood
D	Front/rear plates	1½ × 3½ × 48"	2	SPF
E	Front studs	1½ × 3½ × 81"	4	SPF
F	Door header	1½ × 3½ × 30"	1	SPF
G	Rear studs	1½ × 3½ × 75"	4	SPF
H	Side bottom plate	1½ × 3½ × 30"	2	SPF
I	Top plate	1½ × 3½ × 55"	2	SPF
J	Side front stud	1½ × 3½ × 81"	2	SPF
K	Side middle stud	1½ × 3½ × 71"	2	SPF
L	Side rear stud*	1½ × 3½ × 75¼"	2	SPF
M	Side crosspiece	1½ × 3½ × 27"	2	SPF
N	Door rail (narrow)	¾ × 3½ × 29¾"	1	SPF
O	Door rail (wide)	¾ × 5½ × 23"	2	SPF
P	Door stiles	¾ × 3½ × 71"	2	SPF
Q	Rafters	1½ × 3½ × 44"	4	SPF
R	Outside rafter blocking*	1½ × 3½ × 15¼"	4	SPF
S	Inside rafter blocking*	1½ × 3½ × 18¾"	2	SPF

KEY	PART	DIMENSION	PCS.	MAT.
SIDING & TRIM				
T	Front left panel	¼ × 20 × 85"	1	Siding
U	Front top panel	¼ × 7½ × 30"	1	Siding
V	Front right panel	¼ × 5 × 85"	1	Siding
W	Side panels	¼ × 30½ × 74½"	2	Siding
X	Rear panel	¼ × 48 × 79"	1	Siding
Y	Door panel	¾ × 29¾ × 74"	1	Ext. plywood
Z	Front corner trim	¾ × 3½ × 85"	2	SPF
AA	Front top trim	¾ × 3½ × 50½"	1	SPF
BB	Side casing	¾ × 1½ × 81½"	2	SPF
CC	Top casing	¾ × 1½ × 30"	1	SPF
DD	Bottom casing	¾ × 2½ × 30"	1	SPF
EE	Trim rail (narrow)	¾ × 1½ × 16½"	3	SPF
FF	Trim rail (wide)	¾ × 3½ × 16½"	1	SPF
GG	Side trim	¾ × 2½ × 27"	2	SPF
HH	Side trim	¾ × 2½ × 27¾"	2	SPF
II	Side corner trim (long)	¾ × 1¾ × 85¼"	2	SPF
JJ	Side corner trim (short)	¾ × 1¾ × 79½"	2	SPF
KK	Side trim (wide)	¾ × 3½ × 27"	2	SPF
LL	Side trim (narrow)	¾ × 1½ × 69"	2	SPF
MM	Rear corner trim	¾ × 3½ × 79"	2	SPF
NN	Rear trim (wide)	¾ × 3½ × 50½"	2	SPF
OO	Rear trim (narrow)	¾ × 1½ × 72"	2	SPF
PP	Side windows	¼ × 10 × 28"	2	Acrylic
ROOF				
QQ	Purlins	1½ × 1½ × 61½"	5	
RR	Corrugated closure strips	61½" L	5	
SS	Corrugated roof panels	24 × 46"	3	

*Not shown

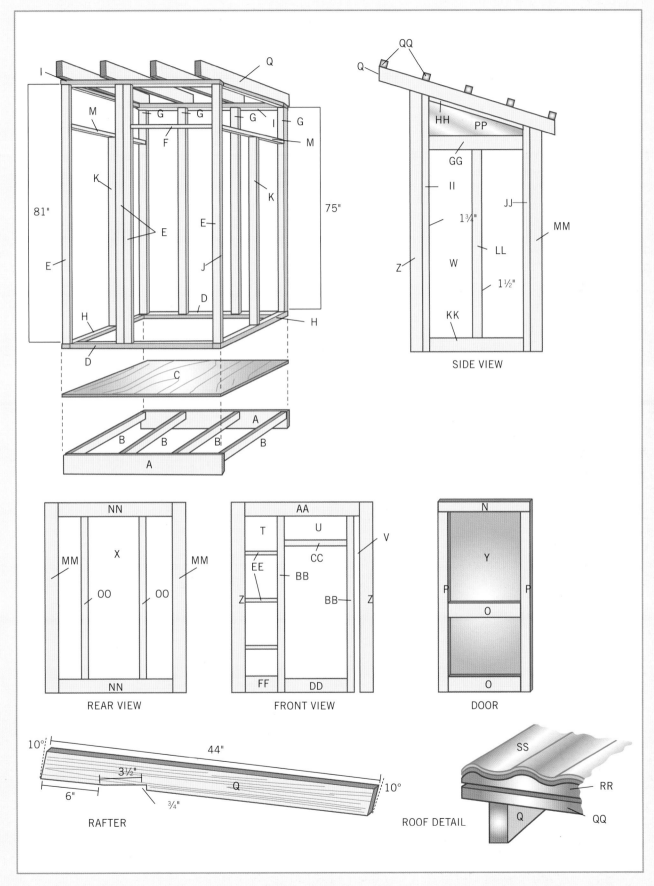

SIDE VIEW

REAR VIEW

FRONT VIEW

DOOR

RAFTER

ROOF DETAIL

How to Build the Mini Garden Shed

Build the Base

Even though moving it is possible, this shed is rather heavy and will require several people or a vehicle to drag it if you build it in your workshop or garage. When possible, determine where you want the shed located and build it in place. Level a 3 × 5-ft. area of ground. The shed base is made of treated lumber, so you can place it directly on the ground. If you desire a harder, more solid foundation, dig down 6" and fill the area with tamped compactible gravel.

Cut the front and rear skids and base crosspieces to length. Place the base parts upside-down on a flat surface and attach the crosspieces to the skids with 2½" deck screws. Working with the parts upside-down makes it easy to align the top edges flush. Cut the base platform to size. Flip the base frame over and attach the base platform (functionally, the floor) with 1½" screws. Set and level the base in position where the shed will be built.

Frame the Shed

Cut the front wall framing members to size, including the top and bottom plates, the front studs,

and the door header. Lay out the front wall on a flat section of ground, such as a driveway or garage floor. Join the wall framing components with 16d common nails (photo 1). Then, cut the rear-wall top and bottom plates and studs to length. Lay out the rear wall on flat ground and assemble the rear wall frame.

Cut both sidewall top and bottom plates to length, and then cut the studs and crosspiece. Miter-cut the ends of the top plate to 10°. Miter-cut the top of the front and rear studs at 10° as well. Lay out and assemble the side walls on the ground. Place one of the side walls on the base platform. Align the outside edge of the wall so it is flush with the outside edge of the base platform. Get a helper to hold the wall plumb while you position the rear wall. If you're working alone, attach a brace to the side of the wall and the platform to hold the wall plumb (photo 2).

Place the rear wall on the platform and attach it to the side wall with 2½" deck screws (photo 3). Align the outside edge of the rear wall with the edge of the platform. Place the front wall on the platform and

Build the wall frames. For the front wall, attach the plates to the outside studs first and then attach the inside studs using the door header as a spacer to position the inside studs.

Raise the walls. Use a scrap of wood as a brace to keep the wall plumb. Attach the brace to the side-wall frame and to the base platform once you have established that the wall is plumb.

Fasten the wall frames. Attach the shed walls to one another and to the base platform with 2½" screws. Use a square and level to check that the walls are plumb and square.

Make the rafters. Cut the workpieces to length, then lay out and cut a birdsmouth notch in the bottom of the two inside rafters. These notches will keep the tops of the inside rafters in line with the outside rafters. The ends should be mitered at 10°.

Install rafter blocking. Some of the rafter blocking must be attached to the rafters by toe-screwing. If you own a pocket screw jig you can use it to drill angled clearance holes for the deck screw heads.

Install the roofing. Attach the corrugated roof panels with 1" neoprene gasket screws (sometimes called pole barn screws) driven through the panels at closure strip locations. Drill slightly oversized pilot holes for each screw and do not overdrive screws—it will compress the closure strips or even cause the panels to crack.

attach it to the side wall with 2½" screws. Place the second side wall on the platform and attach it to the front and rear walls with 2½" screws.

Cut the rafters to length, then miter-cut each end to 10° for making a plumb cut (this way the rafter ends will be perpendicular to the ground). A notch, referred to as a "birdsmouth," must be cut into the bottom edge of the inside rafters so the tops of these rafters align with the outside rafter tops while resting solidly on the wall top plates. Mark the birdsmouth on the inside rafters (see Diagram, page 49) and cut them out with a jigsaw (photo 4). Cut the rafter blocking to length; these fit between the rafters at the front and rear of the shed to close off the area above the top plates. Use the blocking as spacers to position the rafters and then drive 2½" screws up through the top plates and into the rafters. Then, drive 2½" screws through the rafters and into the blocking (photo 5). Toe-screw any rafter blocking that you can't access to fasten through a rafter. Finally, cut the door rails and stiles to length. Attach the rails to the stiles with 2½" screws.

Install the Roofing

This shed features 24"-wide corrugated roofing panels. The panels are installed over wood or foam closure strips that are attached to the tops of 2 × 2 purlins running perpendicular to the rafters. Position the end purlins flush with the ends of the rafters and the inner ones evenly spaced apart along the rafters. The overhang beyond the rafters should be equal on the front and back.

Cut five 61½"-long closure strips. If the closure strips are wood, drill countersunk pilot holes through the closure strips and attach them to the purlins with 1½" screws. Some closure strips are made of foam with a self-adhesive backing. Simply peel off the paper backing and press them in place. If you are installing foam strips that do not have adhesive backing, tack them down with a few pieces of double-sided carpet tape so they don't shift around.

Cut three 44"-long pieces of corrugated roofing panel. Use a jigsaw with a fine-tooth blade or a circular saw with a fine-tooth plywood blade to cut

(continued)

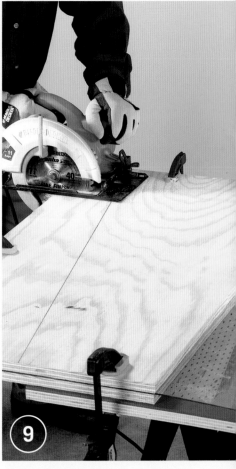

Cut the wall panels. Use a utility knife to score the fiber-cement panel along a straightedge. Place a board under the scored line and then press down on the panel to break the panel as you would with drywall.

Attach siding panels. Attach the fiber-cement siding with 1½" siding nails driven through pilot holes. Space the nails 8 to 12" apart. Drive the nails a minimum ⅜" away from the panel edges and 2" from the corners.

Cut the acrylic window material to size. One way to do this is to sandwich the acrylic between two sheets of scrap plywood and cut all three layers at once with a circular saw (straight cuts) or jigsaw.

fiberglass or plastic panels. Clamp the panels together between scrap boards to minimize vibration while they're being cut (but don't clamp down so hard that you damage the panels). Position the panels over the closure strips, overlapping roughly 4" of each panel and leaving a 1" overhang in the front and back.

Drill pilot holes 12" apart in the field of panels and along the overlapping panel seams. Fasten only in the valleys of the corrugation. The pilot hole diameter should be slightly larger than the diameter of the screw shanks. Fasten the panels to the closure strips and rafters with hex-head screws that are pre-fitted with neoprene gaskets (photo 6, page 51).

Attach the Siding

Cut the siding panels to size by scoring them with a utility knife blade designated for scoring concrete and then snapping them along the scored line (photo 7). Or, use a rented cementboard saw (see page 46). Drill pilot holes in the siding and attach the siding

to the framing with 1½" siding nails spaced at 8 to 12" intervals (photo 8). (You can rent a cementboard coil nailer instead.) Cut the plywood door panel to size. Paint the siding and door before you install the windows and attach the wall and door trim. Apply two coats of exterior latex paint.

Install the Windows

The windows are fabricated from ¼"-thick sheets of clear plastic or acrylic. To cut the individual windows to size, first mark the cut lines on the sheet. To cut acrylic with a circular saw, secure the sheet so that it can't vibrate during cutting. The best way to secure it is to sandwich it between a couple of sheets of scrap plywood and cut through all three sheets (photo 9). Drill ¼"-diameter pilot holes around the perimeter of the window pieces. Position the holes ½" from the edges and 6" apart. Attach the windows to the sidewall framing on the exterior side using 1½" screws (photo 10).

Attach the Trim

Cut the wall and door trim pieces to length. Miter the top end of the side front and rear trim pieces to 10°. Attach the trim to the shed with 2" galvanized finish nails (photo 11). The horizontal side trim overlaps the window and the side siding panel. Be careful not to drive any nails through the plastic window panels. Attach the door trim to the door with 1¼" exterior screws.

Hang the Door

Make the door and fasten a utility handle or gate handle to it. Fasten three door hinges to the door and then fasten the hinges to a stud on the edge of the door opening (photo 12). Use a scrap piece of siding as a spacer under the door to determine the proper door height in the opening. Add hooks and hangers inside the shed as needed to accommodate the items you'll be storing. If you have security concerns, install a hasp and padlock on the mini shed door.

Attach the window panels. Drill a ¼"-dia. pilot hole for each screw that fastens the window panels. These oversized holes give the plastic panel room to expand and contract. The edges of the windows (and the fasteners) will be covered by trim.

Attach the trim boards with 2" galvanized finish nails. In the areas around windows, predrill for the nails so you don't crack the acrylic.

Door

Hang the door using three exterior-rated door hinges. Slip a scrap of ¼"-thick siding underneath the door to raise it off the bottom plate while you install it.

Garbage & Recycling Shed

Trash cans and recycling containers are essential fixtures with any house. If you don't have garage space (or don't want the smells and potential leaks in the garage), you're forced to put these less-than-attractive containers on the outside of your house. No matter where you place them, they don't improve the look of the house or the yard. To make matters worse, area wildlife and even neighborhood dogs find trash cans an irresistible lure. Picking up scattered trash after a family of raccoons has been through the bin is nobody's idea of fun. The answer to all those concerns is, of course, this simple, useful, and handsome shed.

This particular shed has been designed with its single purpose kept clearly in mind, with a design that could not be easier to build. There is no floor or built foundation. The structure is basically a shell that rests on a U-shaped base of pressure-treated 4 × 4s and a façade fabricated from pressure-treated 2 × 4s and durable T1-11 siding. The construction is still rugged enough to protect and secure garbage and recycling wherever you choose to put it—even if it is exposed to the elements. Placed on an existing concrete slab next to the house, a bed of compacted gravel, or on pavers as the shed in this project has been, moisture and freeze-thaw soil movement shouldn't be problems.

The design here is large enough to easily hold a large lidded garbage can and large recycling can side by side. You can scale down the design if you use smaller cans (or add shelves to one bay, if you use bins exclusively for your recycling). In any case, the design includes lids and doors so that you can move the cans out of the shed on trash day, and just pop garbage bags or recyclables into the cans at any other time.

A simple, rugged structure, this shed is meant to be easy to build and utilitarian, while still offering a visual upgrade on bare garbage and recycling cans cluttering your landscape.

SHOPPING LIST

DESCRIPTION	QTY/SIZE	MATERIAL
FOUNDATION		
Drainage material	¾ yard	Compactible gravel
FRAME		
Side bases	2 @ 33½"	4 × 4 pressure treated
Rear base	1 @ 53"	4 × 4 pressure treated
Front posts	2 @ 46½"	2 × 4 pressure treated
Rear posts	4 @ 52½"	2 × 4 pressure treated
Side cross braces	2 @ 33½"	2 × 4 pressure treated
Side top plates	2 @ 38"	2 × 4 pressure treated
Front brace support	2 @ 43"	2 × 4 pressure treated
Front cross brace	2 @ 63"	2 × 4 pressure treated
Front & rear top plates	1 @ 56"	2 × 4 pressure treated
Center partition	1 @ 38" × 52½"	¾" exterior-grade plywood
CLADDING		
Sides	2 @ 38 × 54"	T1-11 siding
Rear	2 @ 31½ × 54"	T1-11 siding
Front edge strip	2 @ 4⅛ × 48"	T1-11 siding
Hinge strip	1 @ 3½ × 63"	T1-11 siding
Front top strip	1 @ 5 × 64¼"	T1-11 siding (2 pieces)

DESCRIPTION	QTY/SIZE	MATERIAL
DOORS & LIDS		
Lids	2 @ 38" × 52½"	⅝" exterior-grade plywood
Doors	2 @ 28 × 46¼"	T1-11 siding
Door stiles	4 @ 41¼"	1 × 3 pine
Door rails	4 @ 23"	1 × 3 pine
HARDWARE		
Door hinges	4 @ 5"	Enameled T hinge
Lid hinges	4 @ 3"	Stainless steel door hinge
Handles	4 @ 3"	Sash handle
Catches	2	Cabinet friction catch
FASTENERS		
6" lag screws	4	
2½" wood screws	30	
3" wood screws	60	
6d galvanized finishing nails	100	
Construction adhesive	1 tube	

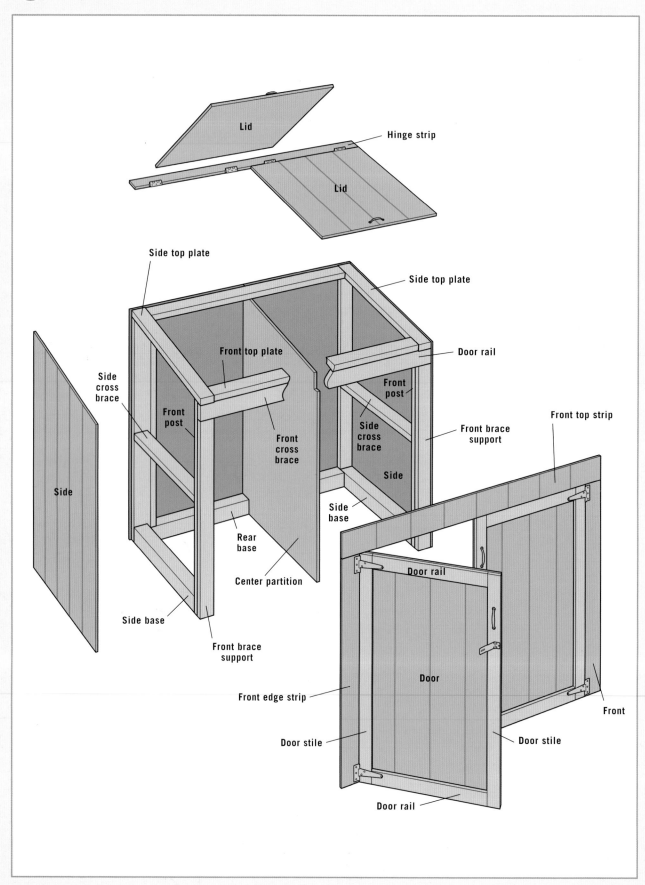

Lid

Hinge strip

Lid

Side top plate

Side top plate

Front top plate

Door rail

Side cross brace

Front post

Front post

Side cross brace

Front top strip

Front cross brace

Side

Front brace support

Side

Side base

Door rail

Front edge strip

Door rail

Rear base

Side base

Center partition

Door

Front

Side base

Front brace support

Door stile

Door stile

Front edge strip

Door

Door stile

Door stile

Door rail

How to Build A Simple Garbage & Recycling Shed

Choose a site for the shed (ideally, as close as possible to the kitchen or back door). If the site is not a concrete slab, it's best to create a stable, level surface for the shed. Here, an existing paver patio was refurbished and leveled for the shed.

Lay out all the sidewall pieces, measure and cut them for both walls. Mark the tops of the front and rear posts on each side for the angle of the top using a framing square.

Make the cuts for all the sidewall pieces, the braces, and top caps. Assemble the side walls. The top caps will be longer than the outside faces of the front and rear posts.

(continued)

Use a speed square to mark the angled cut on the front and back of the sidewall top caps. Make the mitered cuts with a jigsaw.

Cut the front cross brace supports and screw them in place on the front of each sidewall front post, leaving a 3½" space between the top of the support and the bottom of the top plate.

Lay one completed sidewall on top of the 1" plywood for the center partition. Use a 2 × 4 spacer to create a 1½" space between the bottom edge of the wall and the bottom edge of the plywood. Outline the shape of the sidewall onto the plywood, including the angled top cuts and the front and rear edge cuts. Remove the sidewall and cut the partition. Prime and paint or finish the partition, including the edges.

Screw the sidewalls to the rear base using countersunk 6" lag screws. Drive a lag screw through each of the rear wall posts and the side bases.

Stand a rear wall post face to edge against the sidewall rear post so that the back edge of the rear wall post is flush with the back edge of the sidewall rear post. Use the sidewall post as a template to mark the cut line on the top of the rear wall post. Assemble the rear wall by driving 3" screws through the posts, into the ends of the 4 × 4 rear wall base. Screw the rear wall top plate to the posts with 3" wood screws driven down through the top plate and into the end grain of the posts.

Check for square on each side with a framing square. Drive 3" wood screws through the faces of the rear wall posts into the sidewall rear posts' edges, about every 6". Position the front cross brace in place, resting on the front brace supports. Screw the front cross brace to the sidewall front posts with 2½" wood screws.

Mark the centerpoints of the front cross brace and the rear wall 4 × 4 base. Align the plywood partition wall in place with these centerpoints. Check the partition for plumb, and measure to make sure that it is centered perfectly side to side. Drill pilot holes and then nail the rear wall top plate and front cross brace to the edges of the plywood partition. *(continued)*

Position the 56"-long front top plate between the side top plates so the front edge is flush with the front edge of the front cross brace. Screw the front top plate to the front cross brace. Secure the ends of the top plate into each sidewall top plate with screws driven diagonally from below.

Hold one 38" × 55" side-wall T1-11 panel against the sidewall frame with the ripped front and rear edges aligned with the sidewall front and rear edges. (The panel should be about ⅛" from the bottom of the frame to avoid water wicking—use spacers as necessary.) Screw the panel to the side wall, every 6" around the edges and every 10" in the field. Repeat with the opposite sidewall.

Use a jigsaw to cut the siding panels flush with the top face of both sidewall top rails. Nail two ripped-down panels on the rear wall, so the center seam aligns on the partition. Attach 3½" × 48" strips of ripped siding panel on either side of the front, using brads.

Cut the two lids from ⅝" exterior plywood, each 34½" by 31½". Rip the 3½ × 63" hinge strip from the same plywood. Coat the underside of the rear panel with construction adhesive, and use finish nails to fasten the rear panel across the rear top plate and sidewalls.

Paint or finish the lids if applying a top finish. Once dry, attach them to the rear top plate and plywood rear panel with two 3" door hinges per lid. Attach two 3" sash handles centered on the lids at the front edge.

Screw the front frame of 1 × 3 pine pieces around the front edges of the T1-11 door panels, driving the 1" wood screws from the inside of the doors. Mount the doors to the frame of the shed with 5" T hinges, two per door. Install two friction catches for each door, one on either side of the plywood partition, and a slide-bolt latch on the doors.

Call Box Shed

Roughly the size of the rapidly disappearing old-fashioned phone booth or police call box, this shed is just large enough to store a group of long handled yard tools and some bagged supplies. It's also ideally suited for quick and easy modification: add tool hangers on any of the three interior walls, or tack cleats to the sidewalls for one or more shelves.

The shed is designed to sit atop a gravel bed, but it could easily fit right into a discreet corner of a concrete patio. It can even be moved with the aid of one or two helpers, so that you can keep it in different parts of the yard or garden in different seasons, depending on what your needs are any given time of the year.

The door used is an understated siding panel reinforced with a cleat frame on the back. However, you could just as easily put the frame on the front and resize it to the outline of the door for a more conventional, attention-getting look. The design here includes a sliding bolt to keep the shed secure.

Designed roughly to the dimensions of an old-fashioned telephone booth, this shed has a remarkably useful interior space and a handsome appearance that would work well in any yard.

SHOPPING LIST

DESCRIPTION	QTY/SIZE	MATERIAL
FOUNDATION		
Crushed gravel	3 bags	
FRAMING & WALLS		
Skids	2 @ 48"	4 × 4 pressure treated
Floor joists	4 @ 45"	2 × 4 pressure treated
Floor rim joists	2 @ 48"	2 × 4 pressure treated
Floor	1 @ 48" × 48"	⅝" exterior grade
Sidewall studs	8 @ 80"	2 × 4
Front & rear wall studs	10 @ 80"	2 × 4
Front wall jack studs	2 @ 76½"	2 × 4
Front wall header	2 @ 39"	2 × 4
Front & rear top plate	2 @ 48"	2 × 4
Sidewall top plate	2 @ 41"	2 × 4
Sidewall sole plate	2 @ 41"	2 × 4
Rear wall sole plate	1 @ 48"	2 × 4
Front wall sole plate	2 @ 6"	2 × 4
Roof ridge board	1 @ 48"	2 × 4
Gable end stud	2 @ 9¾"	2 × 4
Rafter	8 @ 34¼"	2 × 4
Soffit trim	2 @ 51"	2 × 4
Gable trim	2 @ 48¾"	2 × 4
Roof sheathing	2 @ 35¾" × 51"	2 × 4
Corner trim	8 @ 3½ × 84"	1 × 4 pine

DESCRIPTION	QTY/SIZE	MATERIAL
ROOF		
Drip edge	4 @ 35¾", 2 @ 51"	Galvanized metal
15# roofing paper	36 sq. ft.	
Shingles	⅓ square	Asphalt shingles
SIDING		
Side & rear wall siding	3 @ 48" × 84"	T1-11 siding
Front siding	2 @ 6" × 84", 1 @ 5" × 36"	T1-11 siding
Gable end siding	2 @ 13¼" × 48"	T1-11 siding
DOOR		
Door panel	1 @ 36" × 76½"	T1-11 siding
Frame stile	2 @ 72½"	1 × 3 cedar
Frame rail	3 @ 27"	1 × 3 cedar
FASTENERS & HARDWARE		
Z-flashing	2 @ 48"	
Construction adhesive	1 tube	
Silicone caulk	1 tube	
Door handle	1	5" sash, black
Strap hinges	3	4" black
Bolt lock	1	4"
3" deck screws	20	
2½" deck screws	½ lb.	
1" galvanized roofing nails	½ lb.	
8d galvanized siding nails	½ lb.	
Screening	1 linear foot	Black aluminum

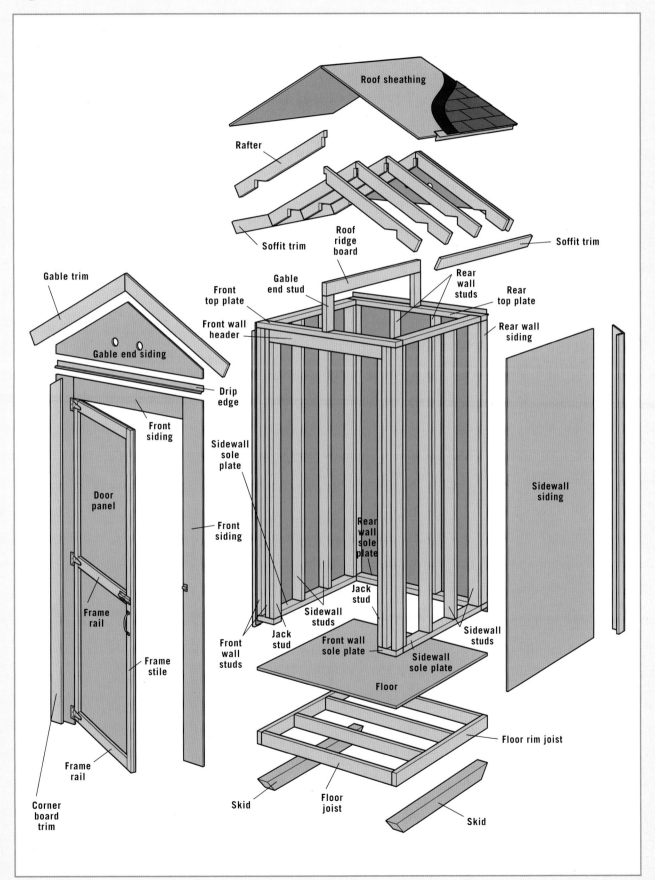

CALL BOX SHED

Roof sheathing

Rafter

Soffit trim

Roof ridge board

Soffit trim

Gable trim

Front top plate

Gable end stud

Rear wall studs

Rear top plate

Front wall header

Rear wall siding

Gable end siding

Drip edge

Front siding

Sidewall siding

Sidewall sole plate

Door panel

Front siding

Rear wall sole plate

Frame rail

Jack stud

Sidewall studs

Sidewall studs

Frame stile

Front wall studs

Jack stud

Front wall sole plate

Sidewall sole plate

Frame rail

Floor

Corner board trim

Floor rim joist

Skid

Floor joist

Skid

How to Build the Call Box Shed

Prepare two parallel 4" deep × 4" wide × 50" long trenches (spaced 44" apart, on center). Fill them with crushed gravel, and compact to a level surface. Check for level between the trenches and along the length of both, using a 4-ft. carpenter's level. Clip each end of the 4 × 4 skids to a 45° angle. Center the skids on the trenches, 48" from one outside face to the other.

On a flat work surface, lay out the floor frame, with the joists and rim joists on edge. Screw the rim joists to the ends of each joist, using 3" deck screws. Measure the diagonals to ensure the frame is square (equal diagonals means the framing is square). Center the floor frame on the skids and toe-screw the joists to the skids. Position the plywood floor on the frame so that all edges are flush, and screw the floor to the frame with 2½" deck screws.

Build the first sidewall, screwing the top and sole plates flush to the outside studs. Screw the inside studs in place, spaced 13¼" on-center. Build the second sidewall in the same way. Brace them in place on the floor, checking to make sure they that they are plumb.

Use the same method to build the rear wall, but with double studs on the edges, and the inside studs spaced 13½" on-center. Position the rear wall on the floor, bracing them to the side walls.

Create the door header by sandwiching two 39" 2 × 4s together with construction adhesive and pairs of 2½" deck screws driven every 4". Center the header on the underside of the top plate and screw it in place using 3" deck screws driven down through the top plate. Screw the king studs to either side of the header, and to the top plate. Install the jack studs on the inside of the king studs. Screw the top plate to the outside front-wall studs, and screw the sole plates to the bottoms of the outside studs, king studs, and jack studs to complete the front wall.

Nail the sidewall sole plates through the floor and into the floor frame. Screw the sidewalls to the rear wall with 3" deck screws. Put the front wall in position, measure the diagonals to ensure square, and screw the sole plates to the floor and the side walls to the front wall.

(continued)

(7)

Starting with the sidewalls, install the plywood siding panels. Side the rear wall, then cut the sections for the front wall and nail them in place.

(8)

Cut the rafters according to the RAFTER PLAN (page 000—cut one and then use it as a template to mark and cut the others). Cut two additional rafters ¾" longer, without the birdmouth cut. These will be the front and back trim pieces.

Toenail the ridge board to the gable studs, and position the ridge board assembly on the top plates, front to back (centered side to side). Toenail it in place, and then check for plumb and adjust as necessary. Toenail the rafters to the top plates, and to the ridge board. The front and rear rafters should be flush with the outer edges of the top plates.

(9)

Install Z-flashing along the top plate on the gable ends, and caulk the edge. Install the gable end plywood panels in place on the front and rear walls, and use a jigsaw to trim off the corners of the panel to match the slope of the end rafters. Use a 3" hole saw to drill two vent holes on either side of the gable end stud.

Use 2½" deck screws to attach the trim boards to the front and rear rafter faces, and along the rafter ends on both sides. Sheath the roof with ½" plywood sections, nailed every 6" along each rafter. Attach the/a metal drip edge along the eaves, and cover the roof with building paper. Then add the/a metal drip edge along the roof edge along both gable ends. Install the asphalt shingles, laying the courses to overlap at the ridge.

Nail the cleat frame to the back of the T1-11 door panel. Hang the door using 3" galvanized strap hinges. Install the handle and sliding bolt. Cut and fasten screening over the inside of the gable siding holes, using construction adhesive. Paint the shed.

Bike Shed

One of the wonderful things about sheds is that they can be designed to just about any size, shape, or configuration, to serve whatever your needs may be. This shed is a perfect example. It was designed to accommodate a family's bikes, and includes a bike hanger on the outside so that the most frequently used two-wheeled convenience can be kept at the ready in good weather.

Like all sheds, this one is also adaptable. If bikes aren't your thing, you can still put it to good use storing sporting equipment, such as plastic youth soccer goals, hockey sticks, balls and bats, and more. You can easily install shelving inside the shed with a couple of cleats made from scrap and nailed to opposing studs, and a plywood shelf cut to fit.

The angled roof with a shorter front wall and taller rear wall lends itself to positioning along a fence or a wall of the house, and the front of the roof includes an extended overhang meant to shield a bike hanging on the outside from rain or other inclement weather. You can add another hanger to store more than one bike, or replace the hangers with a shelf or shelves across the front wall to use as a modest potting bench or shaded area for plants. The door is on the right side of the shed, but if your circumstances require, it could also be installed on the left.

The construction itself has been streamlined—there is no door header or roof ridge board, for instance. The shed is simple and quick to build, with a minimum of time and expertise needed. The pared-back building also means it's easy to customize the shed to the available space. Add a window or vents to turn it into a small potting shed, or even change the footprint measurements to make it a playhouse structure.

This bike shed is designed for ease of construction and storing two-wheeled, manpowered vehicles. But it can just as easily be used to keep tools and building supplies dry and safe.

SHOPPING LIST

DESCRIPTION	QTY/SIZE	MATERIAL
FOUNDATION		
Drainage material	.5 cu. yd.	Compactible gravel
Skids	2 @ 90½"	4 × 4 pressure treated
FLOOR		
Joists	5 @ 38½"	2 × 4 pressure treated
Rim joists	2 @ 90½"	2 × 4 pressure treated
Sheathing	1 sheet @ 4 × 8'	¾" ext. grade plywood
Floor	1 @ 41½ × 90½"	
WALL FRAMING		
Front & rear sole plates	2 @ 90½"	2 × 4
Sidewall sole plate	1 @ 38½"	2 × 4
Studs	10 @ 55¾"	2 × 4
	7 @ 66¾"	2 × 4
Sidewall top plate	1 @ 38½"	2 × 4
Front & rear top plates	2 @ 90½"	2 × 4
ROOF FRAMING		
Rafters	5 @ 56"	2 × 4
Sheathing	2 @ 57½" × 46"	¾" ext. grade plywood
Soffit fascia	1 @ 92"	1 × 4
Rafter trim	2 @ 56¾"	1 × 4

DESCRIPTION	QTY/SIZE	MATERIAL
ROOFING		
Shingles	35 sq. ft.	250# per square min.
15# building paper	35 sq. ft.	
Drip edge	20 linear ft.	Galvanized metal
SIDING		
Wall siding	6 @ 4 × 8'	T1-11 siding (cut to fit)
DOOR		
Panel	1 @ 61½ × 31¼"	T1-11 siding*
Stiles	2 @ 57¼"	1 × 4
Rails	3 @ 24¼"	1 × 4
Door trim	Cut to fit	T1-11 siding
FASTENERS & HARDWARE		
1" galvanized roofing nails	¼ pound	
3" deck screws	½ pound	
2½" deck screws	½ pound	
1½" galvanized wood screws	20	
Surface hinges	3	
Door handle & latch	1	

---

*NOTE: You should be able to cut the door panel from the siding used for the door wall.

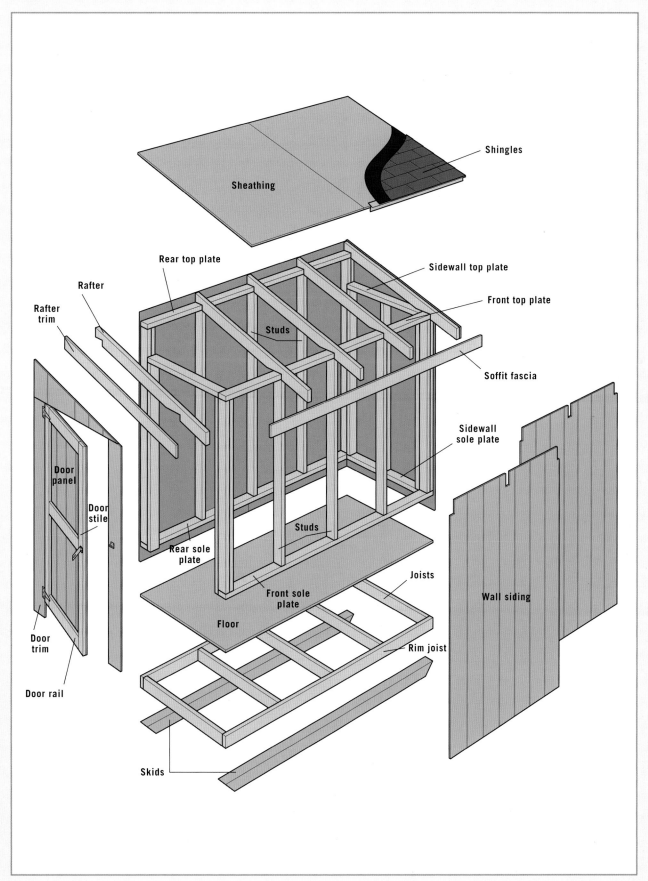

Shingles

Sheathing

Rear top plate

Rafter

Rafter trim

Sidewall top plate

Front top plate

Studs

Soffit fascia

Door panel

Door stile

Sidewall sole plate

Studs

Door trim

Rear sole plate

Joists

Front sole plate

Door rail

Floor

Rim joist

Wall siding

Skids

THE GROUND SCREW ALTERNATIVE

There is a foundation alternative for lighter, simpler buildings such as this bike shed. Ground screws—essentially oversized screws with a top tang sized to hold 4 × 4 posts—are quick and simple to use. They alleviate the need for a gravel base, holding a shed's foundation slightly above grade so moisture isn't a concern. The screws are positioned in the middle and on both ends of rim joists made with 4 × 4 timbers, and then 2 × 4 joists are screwed between the timbers, creating a foundation without the need for skids. To use the screws, you simple mark the locations according to your intended foundation siting, and then screw each one down into the ground, using the supplied rebar handle. The mounting flange on top of the screws includes a series of holes through which lag bolts (also supplied with the screws) are driven to attach the screw to the timber. The screws are made of galvanized steel, so exposure to moisture and the elements won't pose a problem. Although they should not be used in extremely sandy or rocky soils, they are a quick solution in all other types.

These ground screws come with five lag bolts and a rebar turning handle with each screw.

Setting a ground screw is a lot simpler, easier, and quicker than mixing and pouring concrete for a foundation footing or pier.

How to Build the Bike Shed

Prepare the shed site with a 4"-deep bed of compacted gravel. The bed should be longer and wider than the foundation by 1" on all sides. Clip the skid ends to 45 degrees and set in place, parallel, with the outer edges 41½" apart. Level the skids side to side and along their lengths. Assemble the floor frame using 3" deck screws. Toescrew the floor frame to the skids. Screw the floor sheathing down to the floor frame with 3" deck screws, making sure that all edges are flush.

Build the front and rear walls, joining the studs and top and bottom plates with 3" screws. Set the walls in place on the floor with the outer edges of the sole plates flush with the edge of the floor frame. Check to make sure that the walls are plumb, and brace them into position.

Build the left sidewall. Position it between the left ends of the front and rear walls. Screw the walls together using 2½" galvanized wood screws. Check to make sure the construction is square by measuring diagonals.

Screw the door king studs into place against the insides of the front and rear walls, and check for plumb. Toenail the king studs to the floor framing. Screw the top plate to the king studs using 3" deck screws. Remove the front and rear wall braces if you haven't done so already. *(continued)*

Cut a rafter with the two birdmouth cuts and mitered ends. Use the first rafter as a template to mark and cut the remaining rafters. Cut the two 1 × 4 trim pieces to match the rafters (but ¾" longer).

Toenail the rafters in place on the front and rear wall top plates. Attach the trim pieces on each side, and attach the soffit trim to the front (lower) edges of the rafters.

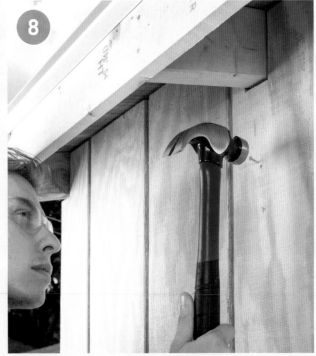

Screw the roof panels to the rafters so the edges are flush with the trim pieces, using 2½" deck screws. Install drip edge along the front (lower) edge of the roof. Cover the roof with building paper, and then install drip edges along the sides and the back. Shingle the roof.

Cut the siding panels to fit the shed walls (the panels on the front must be cut to allow for the rafter ends, so that the siding covers the soffit gaps between rafters). Begin siding with the left side wall, then around the door opening, and finally the front wall and rear wall.

NOTE: The siding extends down over the sides of the floor framing.

Cut the door panel (you should be able to reclaim this from the waste left over from the siding panels for the door-side wall). Check to make sure it fits in the door opening with at least ⅛" gap all the way around, and that the bottom extends down to cover the floor frame when the door is closed.

Screw the stiles to the backside of the door panel, using 1" wood screws. The tops should be flush with the top of the door panel. Screw the rails between the stiles, on the top, bottom, and middle of the door panel.

Position the door in the opening and prop it into place. Mount the door on surface hinges screwed to the front face of the door and siding. Install the latch handle and catch on the door and side wall. Paint or finish the shed, being sure to coat or waterproof the bottom edges of the shed. Because the shed's threshold has a prominent lip, you may want to build a wedge ramp to make it easier to roll bikes into and out of the shed. Follow the steps on page 24 for directions on how to make a ramp for your bike shed. Install a bike hanger on the outside of the front wall of the shed, to keep your bike handy in good weather. (As an alternative, you can install a shelf for potted plants, or individual tool hangers for frequently used garden tools.)

Medium Sheds

Larger yards or bigger families will often benefit by choosing a slightly larger shed. You may sacrifice some portability, but you'll gain quite bit of storage. Opting for a medium-sized shed also means opening up the possible uses for the outbuilding. A structure even slightly larger than a small tool shed can function as a child's playhouse, an outdoor dining area, a workshop, or even a small, quiet home office.

However, choosing a medium-sized shed means paying more attention to visual appeal than if you were just building an enclosure for your garbage cans. A medium shed takes up a significant footprint and can easily be an eyesore if not well appointed or finished attractively.

The upside is that these sheds offer a lot more usable space inside than smaller versions, without a whole lot more work. Pick a style from the following pages, then choose a project and start building!

Turn a medium-sized shed into an overlarge focal point with a few fine details. This yard standout features a hip roof topped by a vent capped in copper, and a large, multi-paned window with detailed faux shutters. It's an extremely eye-pleasing take on a functional structure.

Use distinctive wood to make a medium shed look marvelous. Knotty pine and incredible craftsman details ensure that this simple shed is a real looker. When using softwoods of any kind, make sure they are sealed on all faces and edges to ensure against moisture and insect damage.

Don't hide a medium-sized shed. It's tough to hide anything larger than a tiny tool shed in most landscapes. The wiser strategy is to pick a shed with a crisp look that complements its surroundings. Tucked into a cozy corner and complemented by a handsome fence and stout tree, this country carriage-style shed mimics the classic design of horse-and-carriage houses built alongside larger estates in the 19th century.

Match the shed style to a distinctive yard and garden design. The Japanese teahouse-style shed in this yard reinforces the theme established by the cherry tree, bamboo accents, and fern-heavy landscaping.

Focus on function with a simple, basic kit shed. This roomy metal version is inexpensive and easy to assemble, and comes equipped with flexible storage to accommodate just about any yard-care tool or material you might need to hide away.

Choose prefab that looks custom. The manufacturer of this homey wood shed provides everything you'll need to build a structure that looks completely original. A timber-bordered foundation and charming window box are the finishing touches to a structure with incredible personality that goes along with abundant storage.

Delight children with an unusual mid-sized shed that serves as a whimsical playhouse. Kids will have fun moving between their own porch on their own "home" away from home, and the well-lit interior with a view through two windows.

Dress up your pool house shed. Sprucing up a medium-sized pool house shed takes just a little paint, a small amount of effort, and just about no expertise. A decorative scheme, like the nautical theme and colors used on this shed, provide lots of eye candy for swimmers or sunbathers.

Go bold to supercharge the look of a woodsy shed. The owner of this landscape structure could have chosen to let it blend in with the naturalistic surroundings, but instead added panache with fire-engine red accents and door. It's an upbeat, handsome look that adds to the surroundings.

Let a medium shed shine with a style that doesn't just stand there, but brings it's own pizazz. With a medium-sized shed, it starts to make sense to invest in the actual style, because it's a bigger visual. Here, a modern structure becomes a stunning showpiece in a modern-Japanese style landscape full of focal points.

Indulge a child's whimsy with a medium-sized shed that makes fairy-tale playtime come true. It doesn't take a master craftsman to add the kind of detailing that this shed boasts—just time and attention to detail.

Play it safe with a dialed-back design that doesn't clash. This solid, boxy but handsome shed isn't an eyesore, but it doesn't call attention to itself either. The clean lines and simple craftsmanship are timeless yet easier to build than many other styles.

Create the illusion of a guesthouse to hide a utility shed. There's no better way to conceal rakes, power tools, and bags of manure than in a medium-sized shed that gives no clue as to what's inside. Faux shutters and window boxes complete the convincing illusion that this shed is a small guesthouse rather than a workhorse storage space.

Make a large statement with the right medium-sized getaway shed. If you have the property, a shed can be ideal as a relaxing spot for hiding out or entertaining a few close friends. This medium shed is well outfitted for the occasion, with screened windows, a southern exposure, and a lovely attached deck.

Modern Utility Shed

As its name suggests, this is a useful shed that is well suited to serve many purposes. It's a do-all shed. The 8 × 12-foot footprint provides plenty of room for storage, work space, or a little of both. The 60-inch-wide double doors are wide enough to provide access for a larger lawn tool, such as a riding lawn mower, or they can simply be left open to let in lots of natural light. A 30 × 30-inch window is positioned to the right of the door and is installed high enough to leave room for a workbench to fit below—the perfect place to work on small projects.

This shed features basic wall framing construction. The roof is a classic single-sloped roof. The siding material that we used on this shed is a shiplap 4 × 8 panel called LP SmartSide. These panels are approved to serve as both the sheathing and siding. This type of siding is typically installed with the grooves running vertically, but we chose to install them with the grooves running horizontally. The panels are installed so that the lower panels are overlapped by the shiplap or overlapping edge of the upper panels. The horizontal grooves give this shed a more unique and modern style.

One of the best aspects of this design is that it is easy to modify. If you like the basic design but want a more refined or finished space, there are several ways you could upgrade this shed. For example, you could build it on an elevated framed floor with a wood subfloor so that you could install just about any flooring material. Another option is to replace the plastic window with a manufactured operable window. You could also install just about any finished interior wall covering. The level of finish is limited only by your imagination and your budget.

Work. Store. Play.
This shed does it all—and in a sleek, modern form that will delight everyone.

SHOPPING LIST

DESCRIPTION	QTY/SIZE	MATERIAL
FOUNDATION		
Skids	3 @ 2 × 4 × 12 ft., 10 @ 2 × 4 × 10 ft.	Pressure treated
Floor deck	3 @ ¾ × 4 × 8"	Exterior-grade plywood
WALL FRAMING		
Bottom plates	3 @ 2 × 4 × 8, 1 @ 2 × 4 × 12	Pressure treated
Top plates	2 @ 2 × 4 × 8, 2 @ 2 × 4 × 12	
Front & rear beams	2 @ 4 × 4 × 14	Cedar
Studs	20 @ 2 × 4 × 8, 6 @ 2 × 4 × 12	
Door & window headers, jack studs, cripple studs	4 @ 2 × 4 × 8	
ROOF FRAMING		
Rafters & blocking	12 @ 2 × 8 × 10	Cedar
EXTERIOR FINISHES		
Fascia	2 @ 1 × 8 × 14	
Siding	10 @ ⅜ × 4 × 8	LP Smart Side panels
Window trim	2 @ 1 × 2 × 8	
Corner trim	6 @ 1 × 4 × 8, 2 @ 1 × 4 × 10, 1 @ 1 × 4 × 12	

DESCRIPTION	QTY/SIZE	MATERIAL
ROOFING		
Roofing sheathing	5 @ ¾ × 4 × 8	
15# building paper		
Drip edge	50 linear ft.	
Asphalt shingles	5 bundles	
WINDOWS & SCREENS		
¼"-thick polycarbonate glazing	1 @ 31¼ × 31¼, 3 @ 5¾ × 15¾	
Clear exterior caulk	1 tube	
Window screen	1 roll	
Frames	9 @ 1 × 2 × 8	
FASTENERS & HARDWARE		
J-bolts	12 @ ½"-dia.	
16d common framing nails		
10d common nails		
Box/siding/utility nails	6d × 2"	
Galvanized casing (finish) nails	8d × 2	
1" galvanized roofing nails		
1½" joist hanger nails		
Galvanized rafter (hurricane) straps	14	

SHED FINAL

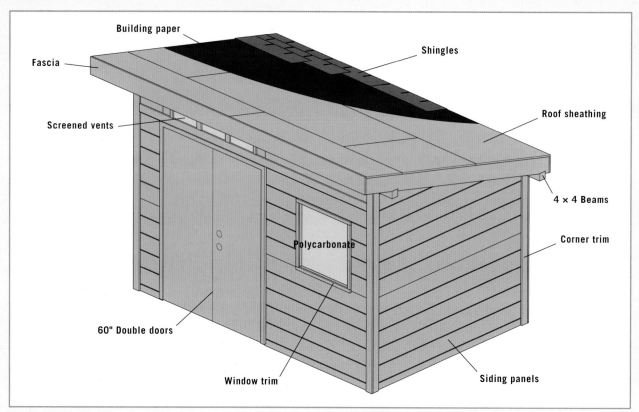

Building paper

Fascia

Shingles

Screened vents

Roof sheathing

4 × 4 Beams

Corner trim

Polycarbonate

60" Double doors

Window trim

Siding panels

FRAMING FULL SHED

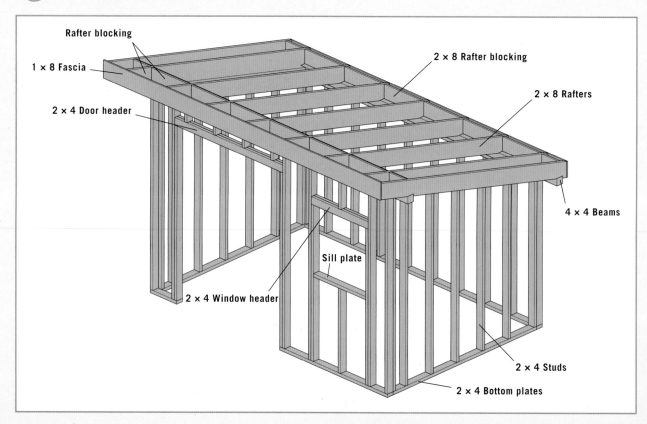

Rafter blocking

1 × 8 Fascia

2 × 8 Rafter blocking

2 × 4 Door header

2 × 8 Rafters

4 × 4 Beams

Sill plate

2 × 4 Window header

2 × 4 Studs

2 × 4 Bottom plates

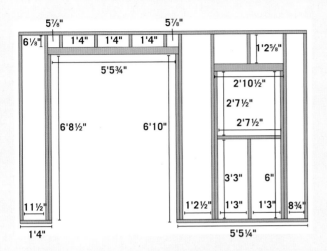

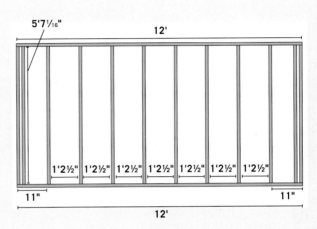

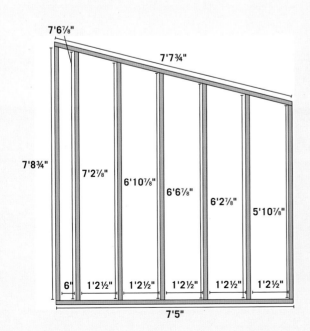

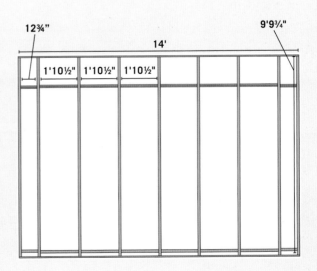

FRONT SHEATHING

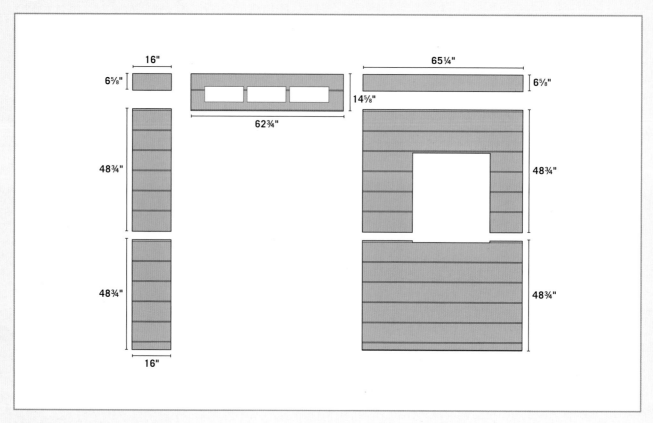

REAR SHEATHING

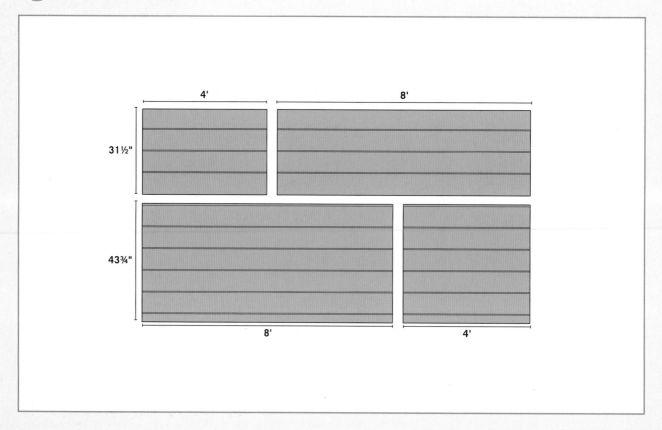

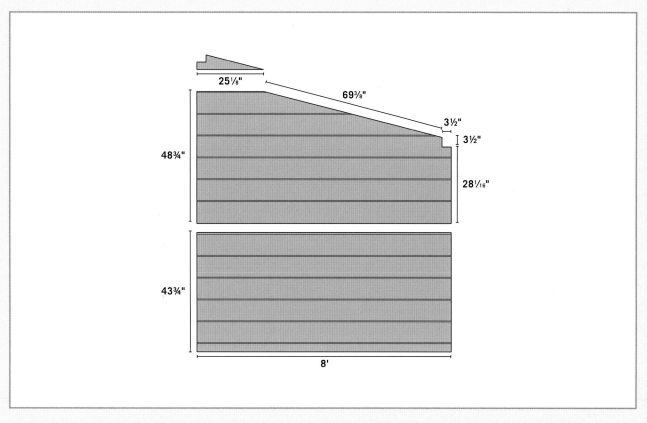

ROOF SHEATHING

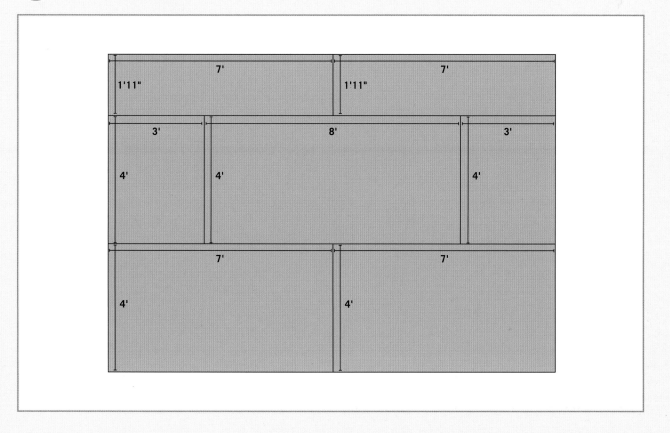

FRONT ELEVATION

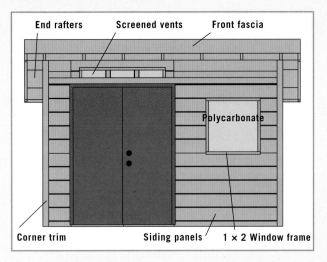

End rafters · Screened vents · Front fascia

Polycarbonate

Corner trim · Siding panels · 1 × 2 Window frame

SIDE ELEVATION

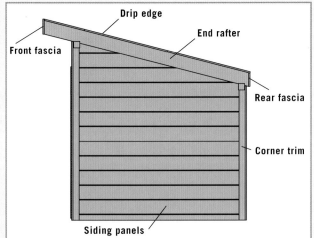

Drip edge

Front fascia · End rafter

Rear fascia

Corner trim

Siding panels

REAR ELEVATION

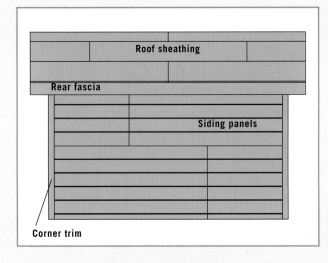

Roof sheathing

Rear fascia

Siding panels

Corner trim

FRONT TRIM DIMENSION

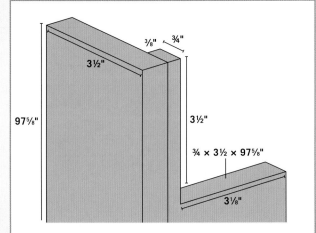

3/8" · 3/4"

3½"

97⅝"

3½"

¾ × 3½ × 97⅝"

3⅛"

REAR TRIM DIMENSION

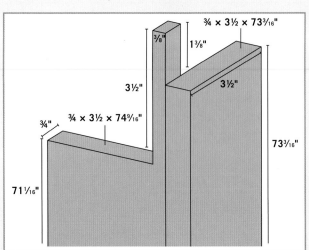

¾ × 3½ × 73³/₁₆"

3/8"

1³/₈"

3½"

3½"

¾ × 3½ × 74⁹/₁₆"

¾"

73³/₁₆"

71¹/₁₆"

SIDE SIDING TOP PIECE DIMENSION

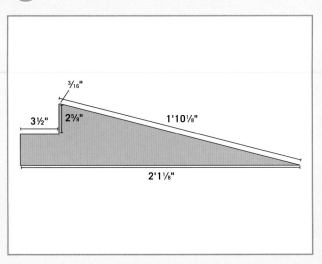

³/₁₆"

3½"

2⅝"

1'10⅛"

2'1⅛"

WINDOW FRAMES

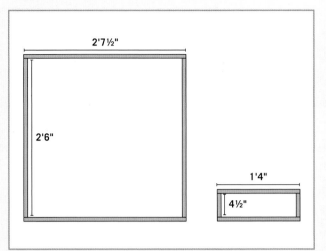

2'7½"

2'6"

1'4"

4½"

FRONT CORNER TRIM DETAIL

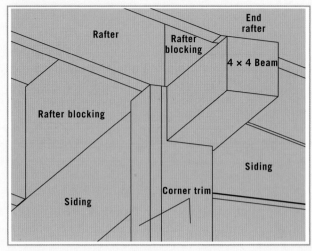

Rafter

Rafter blocking

End rafter

4 × 4 Beam

Rafter blocking

Siding

Siding

Corner trim

RAFTER BLOCKING

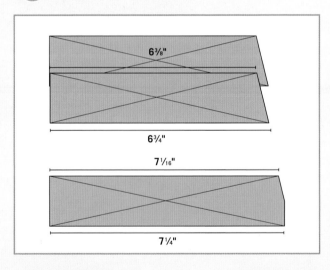

6⅜"

6¾"

7¹⁄₁₆"

7¼"

FLOOR FRAMING

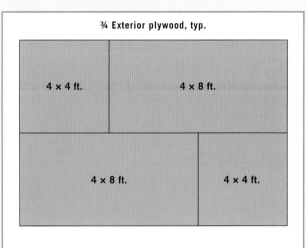

¾ Exterior plywood, typ.

4 × 4 ft.

4 × 8 ft.

4 × 8 ft.

4 × 4 ft.

RAFTER TOP NOTCH

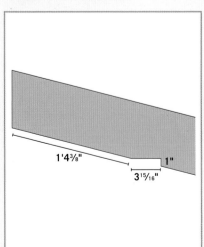

1'4⅜"

1"

3¹⁵⁄₁₆"

END BOTTOM RAFTER NOTCH

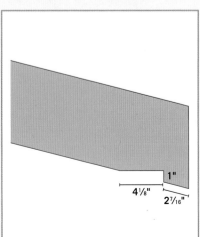

1"

4⅛"

2⁷⁄₁₆"

FIELD BOTTOM RAFTER NOTCH

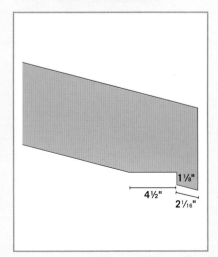

1⅛"

4½"

2¹⁄₁₆"

How to Build the Utility Shed

Build the foundation. Prepare the foundation base by digging a hole matching the shed dimensions, and fill using a 4" layer of compacted gravel. Construct a 2 × 4 floor frame using the FLOOR FRAMING illustration on page 91. Install the frame on gravel and check for level and square, using more gravel under the frame if appropriate.

Attach the sheathing using deck screws, alternating the seams. Try and leave a gap not more than 1¼" between sheets and make sure seams fall above the floor framing members.

Build the rear wall using the REAR WALL FRAMING drawing on page 87. Raise the framed wall into position using a helper. The edges of the sole plate should be flush with the floor sheathing. Level and plumb the frame, and then tack into position using temporary 2 × 4 braces.

Frame both side walls by following the SIDEWALL FRAMING illustration on page 87. Lay out the stud locations on the bottom plate. Cut the studs to length, cutting the top at a 14° angle. Then nail the bottom and top plates to the studs with 16d common nails. Keep the studs perpendicular to the bottom plate and maintain the stud spacing for the full length of each stud.

Attach the side walls to the rear wall using 16d common nails, keeping the outside edges of the walls flush. Plumb the front end of each side wall and then brace them with temporary 2 × 4s attached to the floor foundation.

Frame the front wall by following the FRONT WALL FRAMING illustration on page 87. Attach the front wall to the side walls with 16d common nails. Leave the sidewall bracing in place until you are ready to attach the siding panels.

The sidewall roof overhang is supported by 4 × 4 beams. Attach these beams to the top plate with framing nails. Drive the fasteners up through the top plate and into the beam. If you are not using a pneumatic framing nail gun, then consider attaching the beams with 3" deck screws.

Use a straight edge as a guide and cut the siding panels with a circular saw. The siding panels serve as both the sheathing and siding. These panels are typically used with the long seams running vertically, but in this case we chose to run the long seams horizontally. It is important to orient each panel so that the top piece overlaps the bottom piece. Follow the SIDING SHEATHING illustration on page 89 to cut each piece to the proper width and length. *(continued)*

Attach the siding panels with 6d × 2" ring shank nails. Space nails every six inches around the perimeter and every 12 inches along studs in the field. Carefully align the shiplap evenly with each panel. The vertical edges do not overlap. If you are concerned about water penetration in the vertical seam, then install a piece of double channel (a metal flashing that looks like two pieces of J-channel back-to-back). We chose to leave a 1⁄16" gap between the pieces and fill that gap with exterior-paintable caulk. Though not an acceptable system for a house, it should be sufficient for this shed.

Draw the top angle on the back of a top siding piece with the aid of a helper. Use this same scribing method to mark the irregular pieces that fit around the door and window. Cut along the lines with a circular saw or jigsaw. Paint the shed after all the siding pieces are attached.

Use a carpenter's square and refer to the RAFTER LAYOUT illustrations on page 91 to lay out the rafter notches. There are two different rafter layouts, one for the end rafters and one for the inside (or field) rafters. Use a jigsaw to cut out the rafter notches.

Install blocking pieces on top of the beams between each rafter.

NOTE: Two of the openings between the rafters on the front do not get blocking because the two screen frames are attached in those openings. See the RAFTER BLOCKING illustration on page 91 to cut the rafter blocking to length with a miter saw or circular saw. Then tilt the table saw blade to 14° and bevel cut the top edges. Set the table saw fence to trim 3⁄16" off the front wall blocking. Set the fence to rip the rear wall blocking to 6¾" wide.

Use a table saw to rip the front fascia board to a width of 8¼". The roof sheathing rests on top of the rear fascia board and butts into the front fascia board. The rear fascia board is a full-width 1 × 10, but the front fascia is thinner and must be ripped down from a piece of 1 × 10.

Blade guard removed for clarity

Stain everything at once. Stain the fascia boards, rafters, rafter blocking and all of the stock that you will use for the window/ screen frames and trim boards. Stain all sides of each piece.

NOTE: This is also a good time to stain the bottom sides of the plywood that will be used for the roof sheathing.

Attach the rafters by referring to the RAFTER FRAMING illustration on page 86 to place and tack each rafter in place by toenailing it with 16d common nails. Then secure each rafter with galvanized rafter ties (sometimes referred to as hurricane straps). Attach the rafter ties with galvanized joist hanger nails.

Install the rafter blocking between the rafters. Attach one end of the blocking by driving 10d common nails through the adjacent rafter and into the end of the blocking. Then toenail the other end of the blocking to the next rafter.

(continued)

Make 1 × 2 frames to fit the three openings left between the rafters on the front wall. Then use a staple gun to attach window screen to the back edges of the frames. Trim the excess screen and install the screen frames between the rafters with 2" deck screws. Drill pilot holes through the frame for each screw to prevent splitting the wood.

Attach the front and rear fascia boards to the rafter ends with 8d galvanized casing nails. Then cut the roof sheathing to size using the ROOF SHEATHING LAYOUT illustration on page 89. Stain the underside of the roof sheathing panels if you haven't already. Once the stain is dry, attach the roof sheathing to the rafters with 8d box nails or 2" deck screws. Leave a ⅛" space between each sheet.

Attach the roofing materials. First attach the drip edge along the back edge of the roof. Then attach overlapping layers of building paper to the sheathing with staples. Next, attach drip edge along the side edges and front edge of the roof. Then shingle the roof. Use ⅞" roofing nails so that the nails do not break through the inside face of the sheathing.

Use a circular saw or table saw to cut the plastic for the windows. The best way to cut plastic with a circular saw is to place the sheet of plastic on a large piece of plywood. The plywood supports the plastic and helps prevent it from vibrating during the cut. You will cut into the plywood as you are cutting the plastic. Install a triple-chip tooth blade or plywood blade in your saw and set the blade to a 1" cutting depth. Apply a piece of masking tape over the general cutting line, then mark the actual cutting line on the tape. Clamp a straight piece of wood on the plastic to act as a guide.

The windows are held in place between two frames. Cut the 1 × 2 frame pieces to length following WINDOW FRAME illustrations on page 91. Attach the outside window frame flush with the outside face of the siding with 2" deck screws. Drill pilot holes through the frame for each screw. Deck screws are used to attach the window frames so that they can easily be removed if the window is ever damaged.

Apply a bead of clear exterior caulk along the inside edge of the outside window frame. Set the window in place. Then install the inside window frame pieces. Repeat the window installation process for the screened vents above the door. Next, install the door.

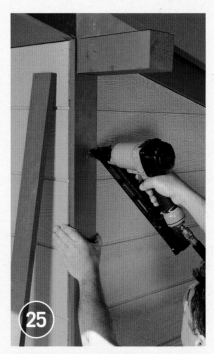

Install the door. For security and ease of access, an outswinging double door made of steel is used in this shed.

Attach the window trim pieces. Measure and cut each piece so that it is flush with the inside edge of the window frames. Attach the trim pieces with 8d galvanized casing nails or pneumatic finish nails.

The side corner trim pieces must be notched to fit around the beam. Follow the CORNER TRIM illustration on pages 90–91. Use a jigsaw to cut the notches. Then attach the corner trim pieces with 8d galvanized casing nails.

Timber-Frame Shed

Timber-framing is a traditional style of building that uses a simple framework of heavy timber posts and beams connected with hand-carved joints. From the outside, a timber-frame building looks like a standard stick-frame structure, but on the inside, the stout, rough-sawn framing members evoke the look and feel of an 18th-century workshop. This 8 × 10-foot shed has the same basic design used in traditional timber-frame structures but with joints that are easy to make.

In addition to the framing, some notable features of this shed are its simplicity and proportions. It's a nicely symmetrical building with full-height walls and an attractively steep-pitched roof, something you seldom find on manufactured kit sheds. The clean styling gives it a traditional, rustic look, but also makes the shed ideal for adding custom details. Install a skylight or windows to brighten the interior, or perhaps cut a crescent moon into the door in the style of old-fashioned backyard privies.

The materials for this project were carefully chosen to enhance the traditional styling. The 1 × 8 tongue-and-groove siding and all exterior trim boards are made from rough-sawn cedar, giving the shed a natural, rustic quality. The door is hand-built from rough cedar boards and includes exposed Z-bracing, a classic outbuilding

detail. As shown here, the roof frame is made with standard 2 × 4s, but if you're willing to pay a little more to improve the appearance, you can use rough-cut 2 × 4s or 4 × 4s for the roof framing.

The timber-frame shed evokes an old-world appeal by using rough-sawn cedar and heavy timber posts.

SHOPPING LIST

DESCRIPTION	QTY/SIZE	MATERIAL
FOUNDATION		
Drainage material	1 cu. yard	Compactible gravel
Skids	3 @ 10'	6 × 6 treated timbers
FLOOR FRAMING		
Rim joists	2 @ 10'	2 × 6 pressure treated
Joists	9 @ 8'	2 × 6 pressure treated
Joist clip angles	18	3 × 3 × 3" × 18-gauge galvanized
Floor sheathing	3 sheets @ 4 × 8'	¾" tongue-&-groove ext.-grade plywood
WALL FRAMING		
Posts	6 @ 8'	4 × 4 rough-sawn cedar
Window posts	2 @ 4'	4 × 4 rough-sawn cedar
Girts	2 @ 10', 2 @ 8'	4 × 4 rough-sawn cedar
Beams	2 @ 10', 2 @ 8'	4 × 6 rough-sawn cedar
Braces	8 @ 2'	4 × 4 rough-sawn cedar
Post bases	6, with nails	Simpson BC40
Post-beam connectors	8 pieces, with nails	Simpson LCE
L-connectors	4, with nails	Simpson A34
Additional posts	6 @ 8'	4 × 4 rough-sawn cedar
ROOF FRAMING		
Rafters	12 @ 7'	2 × 4
Collar ties	2 @ 10'	2 × 4
Ridge board	1 @ 10'	2 × 6
Metal anchors—rafters	8, with nails	Simpson H1
Gable-end blocking	4 @ 7'	2 × 2
EXTERIOR FINISHES		
Siding	2 @ 14', 8 @ 12' 10 @ 10', 29 @ 9'	1 × 8 V-joint rough-sawn cedar
Corner trim	8 @ 9'	1 × 4 rough-sawn cedar
Fascia	4 @ 7', 2 @ 12'	1 × 6 rough-sawn cedar
Fascia trim	4 @ 7', 2 @ 12'	1 × 2 rough-sawn cedar
Subfascia	2 @ 12'	1 × 4 pine
Plywood soffits	1 sheet 4 × 8'	⅜" cedar or fir plywood
Soffit vents (optional)	4 @ 4 × 12"	Louver with bug screen
Flashing (door)	4 linear ft.	Galvanized—18 gauge

DESCRIPTION	QTY/SIZE	MATERIAL
ROOFING		
Roof sheathing	6 sheets @ 4 × 8'	½" ext.-grade plywood
Cedar shingles	1.7 squares	
15# building paper	140 sq. ft.	
Roof vents (optional)	2 units	
DOOR		
Frame	2 @ 7', 1 @ 4'	¾ × 4¼" (actual) S4S cedar
Stops	2 @ 7', 1 @ 4'	1 × 2 S4S cedar
Panel material	7 @ 7'	1 × 6 T&G V-joint rough-sawn cedar
Z-brace	1 @ 8' to 2 @ 8'	1 × 6 rough-sawn cedar
Strap hinges	3	
Trim	5 @ 7'	1 × 3 rough-sawn cedar
Flashing	42" metal flashing	
FASTENERS		
60d common nails	16 nails	
20d common nails	32 nails	
16d galvanized common nails	3½ lbs.	
10d common nails	1 lb.	
10d galvanized casing nails	½ lb.	
8d galvanized box nails	1½ lbs.	
8d galvanized finish nails	7 lbs.	
8d box nails	¼ lb.	
6d galvanized finish nails	40 nails	
3d galvanized finish nails	50 nails	
1½" joist hanger nails	72 nails	
2½" deck screws	25 screws	
1½" wood screws	50 screws	
⅞" galvanized roofing nails	2 lbs.	
⅜" × 6" lag screws, w/ washers	16 screws	
¼" × 6" lag screws, w/ washers		
Construction adhesive	4 tubes	

NOTE: Additional posts may be added as a safety precaution to prevent eave beam deflection.

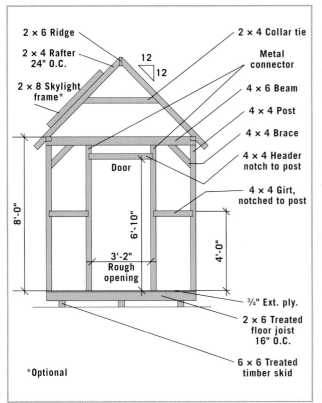

2 × 6 Ridge

2 × 4 Rafter 24" O.C.

2 × 8 Skylight frame*

12 / 12

2 × 4 Collar tie

Metal connector

4 × 6 Beam

4 × 4 Post

4 × 4 Brace

4 × 4 Header notch to post

Door

4 × 4 Girt, notched to post

8'-0"

6'-10"

4'-0"

3'-2" Rough opening

¾" Ext. ply.

2 × 6 Treated floor joist 16" O.C.

6 × 6 Treated timber skid

*Optional

LEFT FRAMING ELEVATION

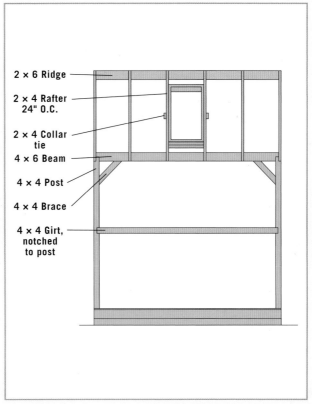

2 × 6 Ridge

2 × 4 Rafter 24" O.C.

2 × 4 Collar tie

4 × 6 Beam

4 × 4 Post

4 × 4 Brace

4 × 4 Girt, notched to post

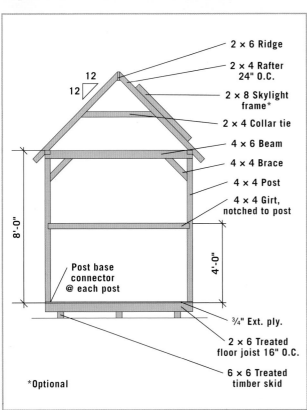

12 / 12

2 × 6 Ridge

2 × 4 Rafter 24" O.C.

2 × 8 Skylight frame*

2 × 4 Collar tie

4 × 6 Beam

4 × 4 Brace

4 × 4 Post

4 × 4 Girt, notched to post

8'-0"

4'-0"

Post base connector @ each post

¾" Ext. ply.

2 × 6 Treated floor joist 16" O.C.

6 × 6 Treated timber skid

*Optional

RIGHT FRAMING ELEVATION

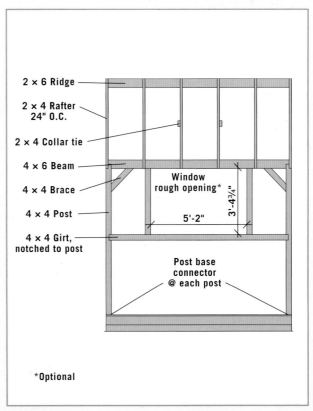

2 × 6 Ridge

2 × 4 Rafter 24" O.C.

2 × 4 Collar tie

4 × 6 Beam

4 × 4 Brace

4 × 4 Post

4 × 4 Girt, notched to post

Window rough opening*

3'-4¾"

5'-2"

Post base connector @ each post

*Optional

BUILDING SECTION

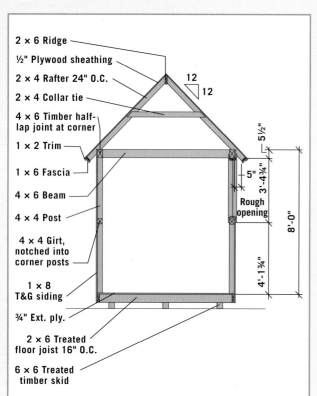

2 × 6 Ridge
½" Plywood sheathing
2 × 4 Rafter 24" O.C.
2 × 4 Collar tie
4 × 6 Timber half-lap joint at corner
1 × 2 Trim
1 × 6 Fascia
4 × 6 Beam
4 × 4 Post
4 × 4 Girt, notched into corner posts
1 × 8 T&G siding
¾" Ext. ply.
2 × 6 Treated floor joist 16" O.C.
6 × 6 Treated timber skid

12
12

5½"
5"
3'-4¾"
Rough opening
8'-0"
4'-1¾"

RAFTER TEMPLATE

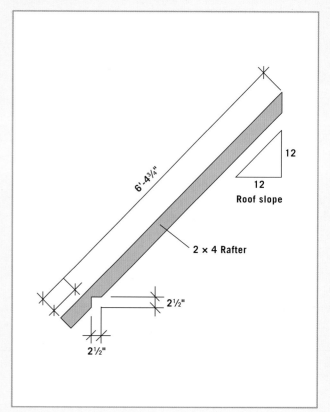

6'-4¾"
12
12
Roof slope
2 × 4 Rafter
2½"
2½"

FLOOR FRAMING PLAN

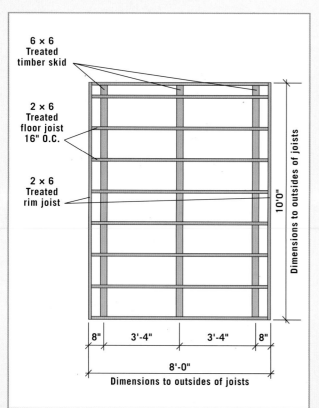

6 × 6 Treated timber skid
2 × 6 Treated floor joist 16" O.C.
2 × 6 Treated rim joist

10'0"
Dimensions to outsides of joists

8"
3'-4"
3'-4"
8"
8'-0"
Dimensions to outsides of joists

FLOOR PLAN

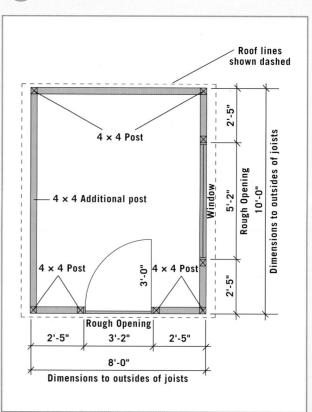

Roof lines shown dashed
4 × 4 Post
4 × 4 Additional post
4 × 4 Post
4 × 4 Post
Window
3'-0"
Rough Opening

2'-5"
5'-2"
Rough Opening
10'-0"
2'-5"
Dimensions to outsides of joists

2'-5"
3'-2"
2'-5"
8'-0"
Dimensions to outsides of joists

FRONT ELEVATION

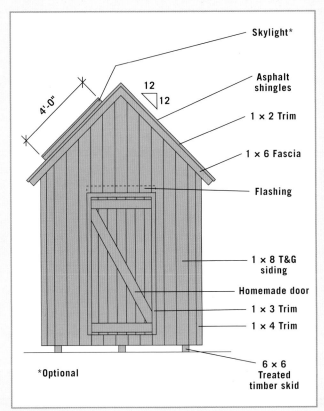

Skylight*

Asphalt shingles

1 × 2 Trim

1 × 6 Fascia

Flashing

12
12

4'-0"

1 × 8 T&G siding

Homemade door

1 × 3 Trim

1 × 4 Trim

6 × 6 Treated timber skid

*Optional

LEFT ELEVATION

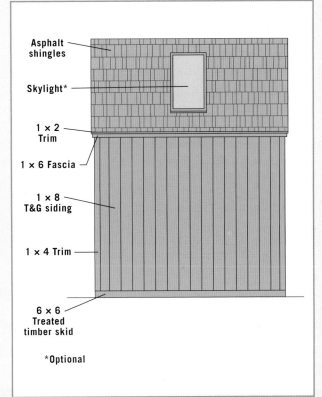

Asphalt shingles

Skylight*

1 × 2 Trim

1 × 6 Fascia

1 × 8 T&G siding

1 × 4 Trim

6 × 6 Treated timber skid

*Optional

REAR ELEVATION

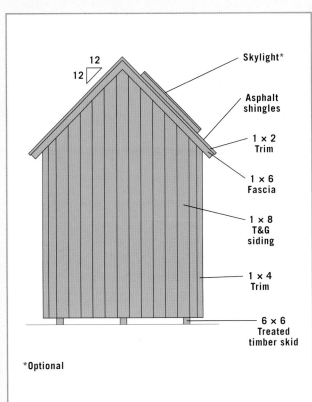

Skylight*

Asphalt shingles

1 × 2 Trim

1 × 6 Fascia

1 × 8 T&G siding

1 × 4 Trim

6 × 6 Treated timber skid

12
12

*Optional

RIGHT ELEVATION

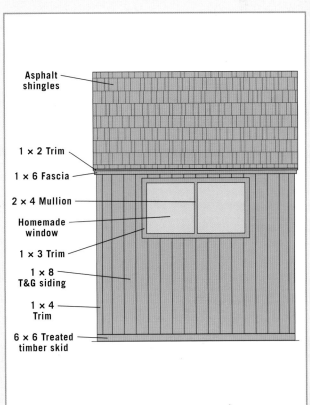

Asphalt shingles

1 × 2 Trim

1 × 6 Fascia

2 × 4 Mullion

Homemade window

1 × 3 Trim

1 × 8 T&G siding

1 × 4 Trim

6 × 6 Treated timber skid

GABLE OVERHANG DETAIL

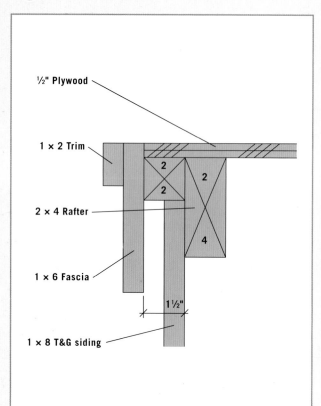

- ½" Plywood
- 1 × 2 Trim
- 2 × 4 Rafter
- 1 × 6 Fascia
- 1½"
- 1 × 8 T&G siding

2
2
2
4

EAVE DETAIL

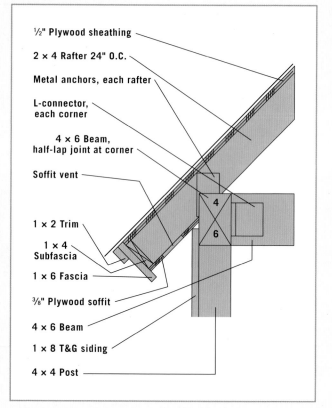

- ½" Plywood sheathing
- 2 × 4 Rafter 24" O.C.
- Metal anchors, each rafter
- L-connector, each corner
- 4 × 6 Beam, half-lap joint at corner
- Soffit vent
- 1 × 2 Trim
- 1 × 4 Subfascia
- 1 × 6 Fascia
- ⅜" Plywood soffit
- 4 × 6 Beam
- 1 × 8 T&G siding
- 4 × 4 Post

4
6

DOOR JAMB DETAIL

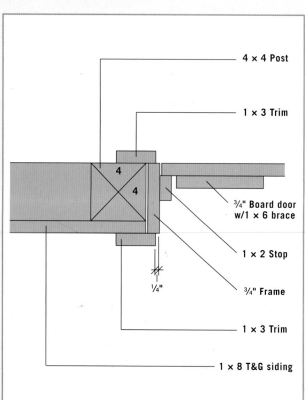

- 4 × 4 Post
- 1 × 3 Trim
- ¾" Board door w/1 × 6 brace
- 1 × 2 Stop
- ¾" Frame
- 1 × 3 Trim
- 1 × 8 T&G siding

4
4

¼"

DOOR DETAIL

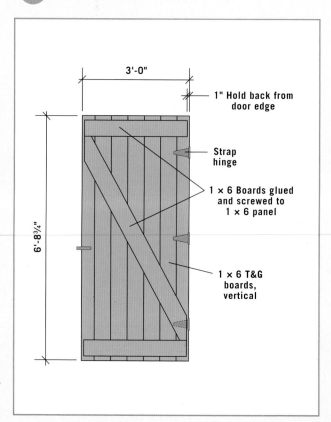

- 3'-0"
- 1" Hold back from door edge
- Strap hinge
- 1 × 6 Boards glued and screwed to 1 × 6 panel
- 1 × 6 T&G boards, vertical
- 6'-8¾"

How to Build the Timber-Frame Shed

Prepare the foundation site with a 4"-deep layer of compacted and leveled gravel. Cut three 6 × 6 treated timber skids (120"). Place the skids following the FLOOR FRAMING PLAN (page 102). Lay a straight 2 × 4 across the skids and test with a level.

Cut two 2 × 6 rim joists (120") and nine joists (93"). Assemble the floor frame with galvanized nails, as shown in the FLOOR FRAMING PLAN. Check the frame to make sure it is square by measuring the diagonals.

Install joist clip angles at each joist along the two outer skids with galvanized nails. Toenail each joist to the center skid.

Install the tongue-and-groove plywood floor sheathing, starting with a full sheet at one corner of the frame. The flooring should extend all the way to the outside edges of the floor frame. *(continued)*

To prepare the wall posts, cut six 4 × 4 posts (90½"), making sure both ends are square. On the four corner posts, mark for 3½"-long × 1½"-deep notches (to accept the girts) on the two adjacent inside faces of each post. Start the notches 46¼" from the bottom ends of the posts.

Mark the door frame posts for notches to receive a girt at 46¼" and for the door header at 82"; see FRONT FRAMING ELEVATION (page 101). Remove the waste from the notch areas with a circular saw and clean up with a broad wood chisel. Test-fit the notches to make sure the 4 × 4 girts will fit snugly.

Position the post bases so the posts will be flush with the outsides of the shed floor. Install the bases with 16d galvanized common nails. The insides of the door posts should be 29" from the floor sides. Brace each post so it is perfectly plumb, and then fasten it to its base using the base manufacturer's recommended fasteners.

Cut two 4 × 6 beams at 10 ft. and two at 8 ft. Notch the ends of the beams for half-lap joints: Measure the width and depth of the beams and mark notches equal to the width × ½ the depth. Orient the notches as shown in the FRAMING ELEVATIONS (page 101). Cut the notches with a handsaw, then test-fit the joints, and make fine adjustments with a chisel.

Set an 8-ft. beam onto the front wall posts and tack it in place with a 16d nail at each end. Tack the other 8-ft. beam to the back posts. Then, position the 10 ft. beams on top of the short beam ends, forming the half-lap joints. Measure the diagonals of the front wall frame to make sure it's square, and then anchor the beams with two 60d galvanized nails at each corner (drill pilot holes for the nails).

Reinforce the beam connections with a metal post-beam connector on the outside of each corner and on both sides of the door posts, using the recommended fasteners. Install an L-connector on the inside of the beam-to-beam joints; see the EAVE DETAIL (page 104).

Cut eight 4 × 4 corner braces (20"), mitering the ends at 45°. Install the braces flush with the outsides of the beams and corner posts, using two ⅜ × 6" lag screws (with washers) driven through counterbored pilot holes.

Measure between the posts at the notches, and cut the 4 × 4 girts to fit. To allow the girts to meet at the corner posts, make a 1½ × 1½" notch at both ends of the rear wall girts and the outside ends of the front wall girts. Install the girts with construction adhesive and two 20d nails driven through the outsides of the posts (through pilot holes). Cut and install the 4 × 4 door header in the same fashion. *(continued)*

Frame the roof: Cut two pattern rafters using the RAFTER TEMPLATE (page 102). Test-fit the patterns, and then cut the remaining ten rafters. Cut the 2 × 6 ridge (120"). Install the rafters and ridge using 24" on-center spacing. Cut four 2 × 2s to extend from the roof peak to the rafter ends, and install them flush with the tops of the rafters; see the GABLE OVERHANG DETAIL (page 104). Add framing connectors at the rafter-beam connections (except the outer rafters).

NOTE: If desired, you can add framing for a skylight.

Cut four 2 × 4 collar ties (58"), mitering the tops of the ends at 45°. Install the ties ½" below the tops of the rafters, as shown in the FRAMING ELEVATIONS (page 101).

Install the 1 × 8 siding on the front and rear walls so it runs from the 2 × 2s down to ¾" below the bottom of the floor frame. Fasten the siding with 8d corrosion-resistant finish nails or siding nails. Don't nail the siding to the door header in this step.

Cover the rafter ends along the eaves with 1 × 4 subfascia, flush with the tops of the rafters; see the EAVE DETAIL (page 104). Install the 1 × 6 fascia and 1 × 2 trim at the gable ends, then along the eaves, mitering the corner joints. Keep the fascia and trim ½" above the rafters so it will be flush with the roof sheathing.

Rip the plywood soffit panels to fit between the wall framing and the fascia, and install them with 3d galvanized box nails; see the EAVE DETAIL (page 104).

Deck the roof with ½" plywood sheathing, starting at the bottom corners. Cover the sheathing with building paper, overhanging the 1 × 2 fascia trim by ¾". Install the cedar shingle roofing or asphalt shingles. Include roof vents, if desired (they're a good idea). Finish the roof at the peak with a 1× ridge cap.

Construct the door frame from ¾" × 4¼" stock. Cut the head jamb at 37¾" and the side jambs at 81". Fasten the head jamb over the ends of the side jambs with 2½" deck screws. Install the frame in the door opening, using shims and 10d galvanized casing nails. Add 1 × 2 stops to the jambs, ¾" from the outside edges.

Build the door with seven pieces of 1 × 6 siding cut at 80¾". Fit the boards together, then mark and trim the outer pieces so the door is 36" wide. Install the 1 × 6 Z-bracing with adhesive and 1¼" wood screws, as shown in the DOOR DETAIL (page 104). Install flashing over the outside of the door, then add 1 × 3 trim around both sides of the door opening, as shown in the DOOR JAMB DETAIL (page 104). Hang the door with three strap hinges.

Gothic Playhouse

Playhouses are all about stirring the imagination. Loaded with fancy American Gothic details, this charming little house makes a special play home for kids and an attractive backyard feature for adults. In addition to its architectural character (see Gothic Style, below), what makes this a great playhouse design is its size—the enclosed house measures 5 × 7½ feet and includes a 5-foot-tall door and plenty of headroom inside. This means your kids will likely "outgrow" the playhouse before they get too big for it. And you can always give the house a second life as a storage shed.

At the front of the house is a 30-inch-deep porch complete with real decking boards and a nicely decorated railing. Each side wall features a window and flower box, and the "foundation" has the look of stone created by wood blocks applied to the floor framing. All of these features are optional, but each one adds to the charm of this well-appointed playhouse.

As shown here, the floor of the playhouse is anchored to four 4 × 4 posts buried in the ground.

As an alternative, you can set the playhouse on 4 × 6 timber skids. Another custom variation you might consider is in the styling of the verge boards (the gingerbread gable trim). Instead of using the provided pattern, you can create a cardboard template of your own design. Architectural plan and pattern books from the Gothic period are full of inspiration for decorative ideas.

GOTHIC STYLE

The architectural style known as American Gothic (also called Gothic Revival and Carpenter Gothic) dates back to the 1830s and essentially marks the beginning of the Victorian period in American home design. Adapted from a similar movement in England, gothic style was inspired by the ornately decorated stone cathedrals found throughout Europe. The style quickly evolved in America as thrifty carpenters learned to re-create and reinterpret the original decorative motifs using wood instead of stone.

American Gothic's most characteristic feature is the steeply pitched roof with fancy scroll-cut bargeboards, or verge boards, which gave the style its popular nickname, "gingerbread." Other typical features found on gothic homes (and the gothic playhouse) include board-and-batten siding, doors and windows shaped with gothic arches, and spires or finials adorning roof peaks.

Hosting tea with the Mad Hatter or fending off dragons around the gables, the gothic playhouse is all about sparking the imagination.

SHOPPING LIST

DESCRIPTION	QTY/SIZE	MATERIAL
FOUNDATION/FLOOR		
Drainage material	1 cu. yd.	Compactible gravel
Foundation posts	4 @ field measure	4 × 4 pressure-treated landscape timbers
Concrete	Field measure	3,000 psi concrete
Rim joists	3 @ 10', 1 @ 8'	2 × 12 pressure treated, rated for ground contact
Floor joists	1 @ 10', 2 @ 8'	2 × 6 pressure treated
Box sills (rim joists)	2 @ 12'	2 × 4 pressure treated
Floor sheathing	2 sheets @ 4 × 8'	¾" ext.-grade plywood
Porch decking	5 @ 10'	1 × 6 pressure treated decking
Foundation "stones"	7 @ 10'	5/4 × 6" treated decking w/ radius edge (R.E.D.), rated for ground contact
FRAMING		
Wall framing & railings	29 @ 12'	2 × 4
Rafters & spacers	7 @ 12'	2 × 4
Ridge board	1 @ 8'	1 × 6
Collar ties	1 @ 10'	1 × 4
EXTERIOR FINISHES		
Siding, window boxes & door trim	26 @ 10'	1 × 8 pressure treated or cedar
Battens & trim	30 @ 8'	1 × 2 pressure treated or cedar
Door panel, verge boards & fascia	10 @ 10'	1 × 6 pressure treated or cedar
Door braces, trim & railing trim	2 @ 10'	1 × 4 pressure treated or cedar
Railing balusters	4 @ 8'	2 × 2 pressure treated or cedar
Window stops	2 @ 8'	⅜" pressure-treated or cedar quarter-round molding
Window glazing (optional)	4 @ 20 × 9½"	¼" plastic glazing

DESCRIPTION	QTY/SIZE	MATERIAL
SPIRE		
Post	1 @ 8'	4 × 4 pressure treated
Trim	1 @ 4'	1 × 2 pressure treated
Molding	1 @ 4'	Cap molding, pressure treated
Balls	2 @ 3"-dia.	Wooden sphere, pressure treated
ROOFING		
Sheathing	4 sheets @ 4 × 8'	½" exterior-grade plywood roof sheathing
15# building paper	1 roll	
Drip edge	40 linear ft.	Metal drip edge
Shingles	1 square	Asphalt shingles— 250# per sq. min.
FASTENERS & HARDWARE		
16d galvanized common nails	3½ lbs.	
16d common nails	5 lbs.	
10d common nails (for double top plates)	½ lb.	
10d galvanized finish/ casing nails	4 lbs.	
8d galvanized common nails	1 lb.	
8d box nails	2 lbs.	
8d galvanized siding nails	8 lbs.	
1" galvanized roofing nails	3 lbs.	
2" deck screws (for porch decking)	1 lb.	
6d galvanized finish nails	2 lbs.	
3½" galvanized wood screws	24 screws	
1¼" galvanized wood screws	12 screws	
Dowel screws (for spire)	3 screws	Galvanized dowel screws
Lag screws w/washers	2 @ 6"	½" galvanized lag screws
Door hinges w/screws	3	Corrosion-resistant hinges
Door handle/latch	1	
Exterior wood glue		
Clear exterior caulk (for optional window panes)		
Construction adhesive		

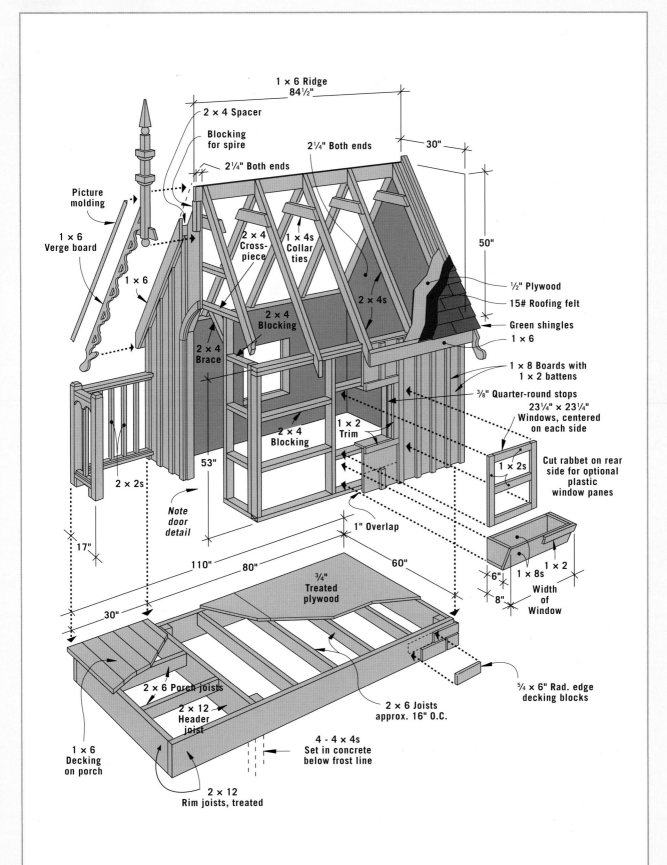

1 × 6 Ridge
84½"

2 × 4 Spacer

Blocking
for spire

2¼" Both ends

30"

2¼" Both ends

Picture
molding

1 × 6
Verge board

1 × 6

2 × 4
Cross-
piece

1 × 4s
Collar
ties

50"

2 × 4s

½" Plywood

15# Roofing felt

Green shingles

1 × 6

2 × 4
Blocking

2 × 4
Brace

1 × 8 Boards with
1 × 2 battens

⅜" Quarter-round stops

23¼" × 23¼"
Windows, centered
on each side

2 × 4
Blocking

1 × 2
Trim

1 × 2s

Cut rabbet on rear
side for optional
plastic
window panes

53"

2 × 2s

Note
door
detail

1" Overlap

1 × 2

1 × 8s

6"

8"

Width
of
Window

17"

110"

80"

60"

¾"
Treated
plywood

30"

5/4 × 6" Rad. edge
decking blocks

2 × 6 Porch joists

2 × 12
Header
joist

2 × 6 Joists
approx. 16" O.C.

1 × 6
Decking
on porch

4 - 4 × 4s
Set in concrete
below frost line

2 × 12
Rim joists, treated

FLOOR PLAN

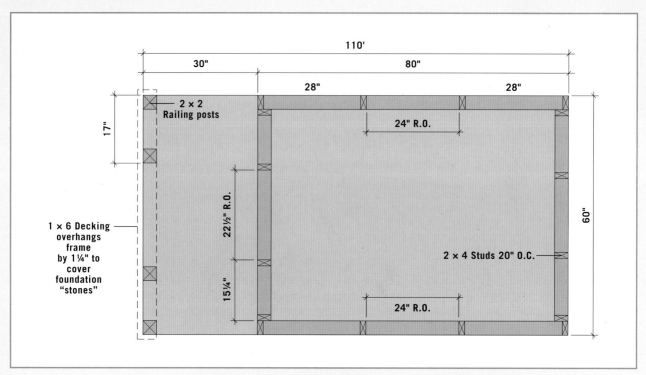

110'

30" | 80"

28" | 28"

2 × 2
Railing posts

17"

24" R.O.

22½" R.O.

15¼"

1 × 6 Decking
overhangs
frame
by 1¼" to
cover
foundation
"stones"

60"

2 × 4 Studs 20" O.C.

24" R.O.

VERGE BOARD TEMPLATE

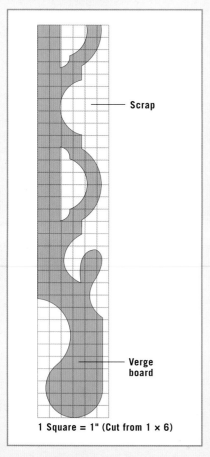

Scrap

Verge
board

1 Square = 1" (Cut from 1 × 6)

DECK RAILING DETAIL

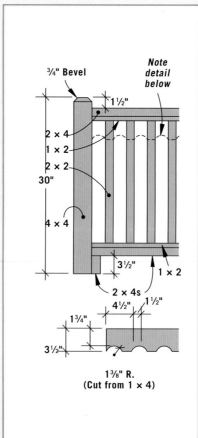

¾" Bevel

*Note
detail
below*

1½"

2 × 4

1 × 2

2 × 2

30"

4 × 4

3½"

1 × 2

2 × 4s

4½" | 1½"

1¾"

3½"

1⅜" R.
(Cut from 1 × 4)

SPIRE DETAIL

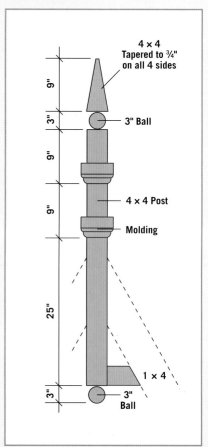

4 × 4
Tapered to ¾"
on all 4 sides

9"

3"   3" Ball

9"

9"   4 × 4 Post

Molding

25"

1 × 4

3"   3"
        Ball

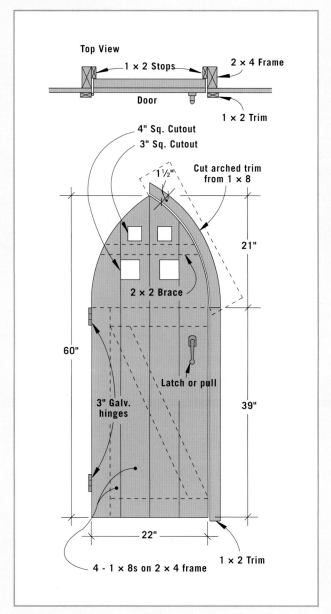

Top View

1 × 2 Stops

2 × 4 Frame

Door

1 × 2 Trim

4" Sq. Cutout

3" Sq. Cutout

Cut arched trim from 1 × 8

1½"

21"

2 × 2 Brace

60"

Latch or pull

39"

3" Galv. hinges

22"

1 × 2 Trim

4 - 1 × 8s on 2 × 4 frame

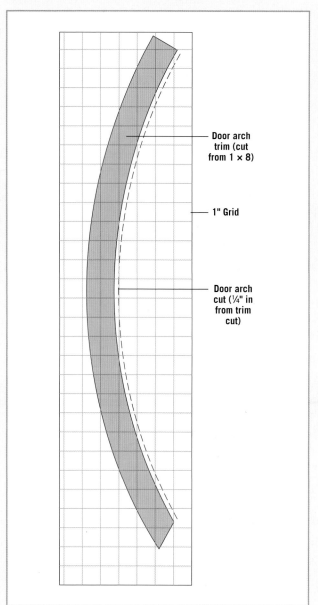

Door arch trim (cut from 1 × 8)

1" Grid

Door arch cut (¼" in from trim cut)

 **BOARD & BATTEN DETAIL**

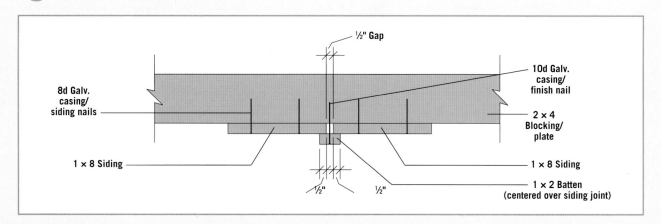

½" Gap

10d Galv. casing/ finish nail

8d Galv. casing/ siding nails

2 × 4 Blocking/ plate

1 × 8 Siding

1 × 8 Siding

1 × 2 Batten (centered over siding joint)

½"

½"

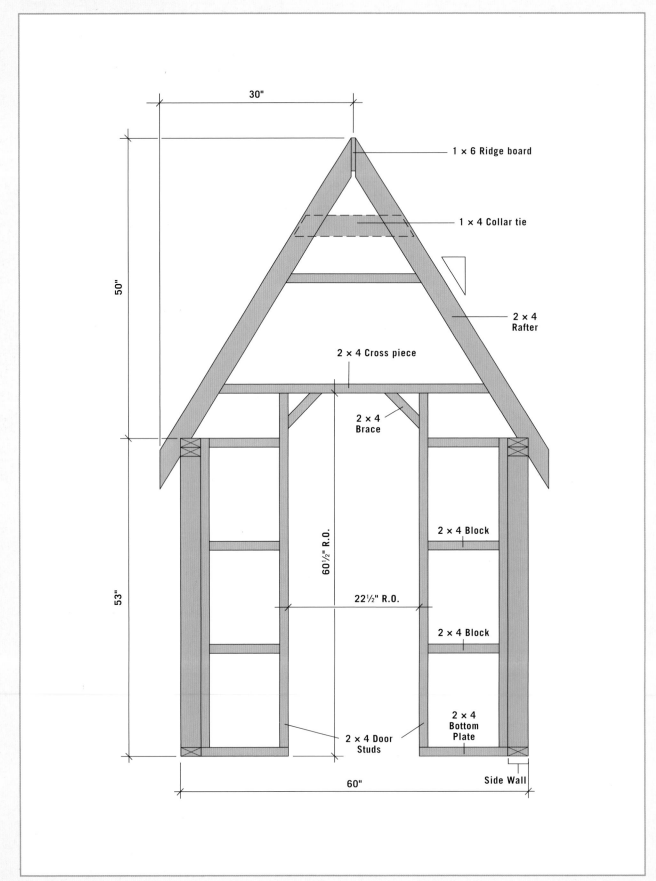

30"

1 × 6 Ridge board

1 × 4 Collar tie

2 × 4 Rafter

50"

2 × 4 Cross piece

2 × 4 Brace

2 × 4 Block

60½" R.O.

53"

22½" R.O.

2 × 4 Block

2 × 4 Bottom Plate

2 × 4 Door Studs

60"

Side Wall

FLOOR FRAMING PLAN

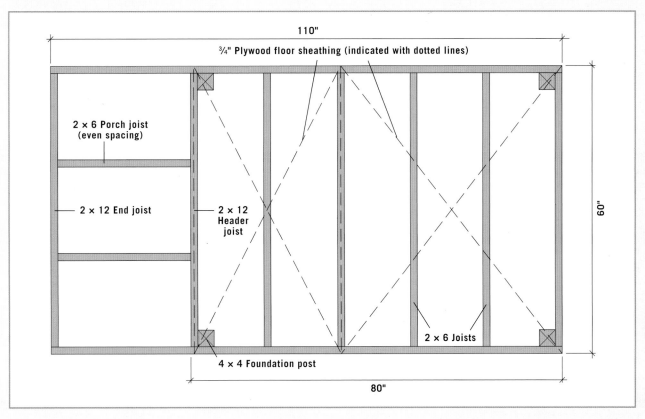

110"

¾" Plywood floor sheathing (indicated with dotted lines)

2 × 6 Porch joist
(even spacing)

2 × 12 End joist

2 × 12
Header
joist

60"

2 × 6 Joists

4 × 4 Foundation post

80"

SIDE FRAMING

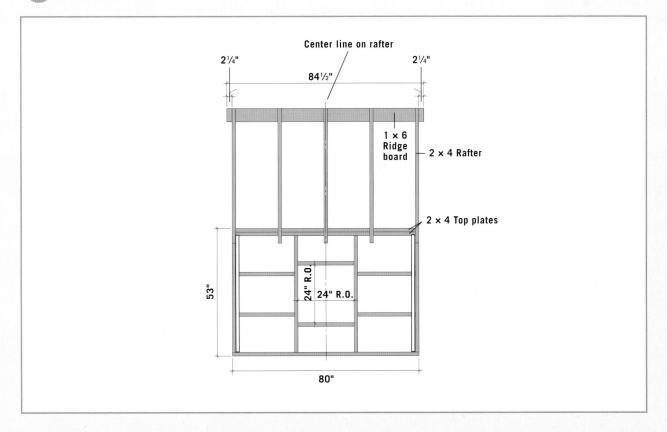

Center line on rafter

2¼"

84½"

2¼"

1 × 6
Ridge
board

2 × 4 Rafter

2 × 4 Top plates

53"

24" R.O.

24" R.O.

80"

RAFTER TEMPLATE

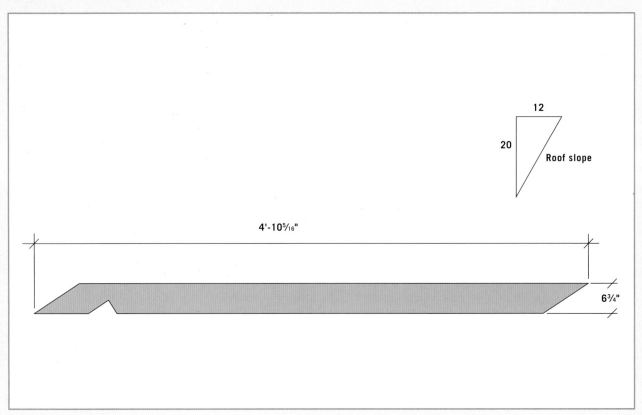

12

20

Roof slope

4'-10⁵⁄₁₆"

6³⁄₄"

WINDOW BOX DETAIL

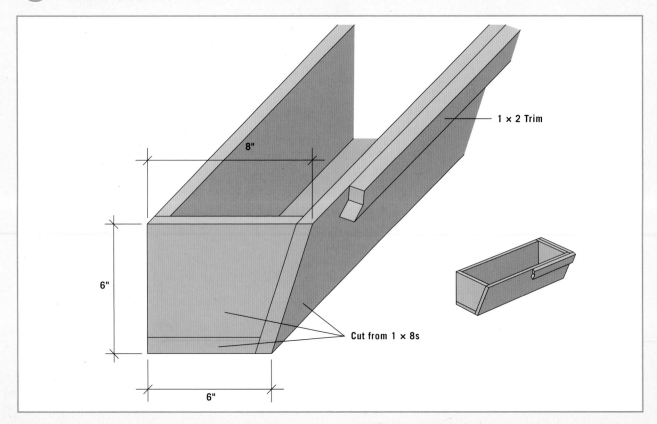

1 × 2 Trim

8"

6"

6"

Cut from 1 × 8s

How to Build the Gothic Playhouse

①

②

Set up perpendicular mason's lines and batter boards to plot out the excavation area and the post hole locations, as shown in FLOOR FRAMING PLAN (page 117). Excavate and grade the construction area, preparing for a 4"-thick gravel base. Dig 12"-dia. holes to a depth below the frost line, plus 4". Add 4" of gravel to each hole. Set the posts in concrete so they extend about 10" above the ground.

After the concrete dries add compactible gravel and tamp it down so it is 4" thick and flat. Cut two 2 × 12 rim joists for the floor frame, two 2 × 12 end joists and one header joist. Cut four 2 × 6 joists at 57" and two porch joists at 27¾". Assemble the floor frame with 16d galvanized common nails following FLOOR FRAMING PLAN (page 117).

③

④

Make sure the frame is square and level (prop it up temporarily), and then fasten it to the posts with 16d galvanized common nails.

Cover the interior floor with plywood, starting at the rear end. Trim the second piece so it covers ½ of the header joist. Install the 1 × 6 porch decking starting at the front edge and leaving a ⅛" gap between boards. Extend the porch decking 1¼" beyond the front and sides of the floor frame. *(continued)*

Frame the side walls as shown in SIDE FRAMING (page 117) and the FLOOR PLAN (page 114). Each wall has four 2 × 4 studs at 48½", a top and bottom plate at 80", and a 2 × 4 window header and sill at 24". Install the horizontal 2 × 4 blocking, spaced evenly between the plates. Install only one top plate per wall at this time.

Build the rear wall. Raise the side and rear walls, and fasten them to each other and to the floor frame. Add double top plates. Both sidewall top plates should stop flush with the end stud at the front of the wall.

To frame the front wall, cut two treated bottom plates at 15¼", two end studs at 51½", and two door studs at 59". Cut a 2 × 4 crosspiece and two braces, mitering the brace ends at 45°. Cut six 2 × 4 blocks at 12¼". Assemble the wall as shown in FRONT FRAMING (page 116). Raise the front wall and fasten it to the floor and sidewall frames.

Cut one set of 2 × 4 pattern rafters following the RAFTER TEMPLATE (page 118). Test-fit the rafters and make any necessary adjustments. Use one of the pattern rafters to mark and cut the remaining eight rafters. Also cut four 2 × 4 spacers—these should match the rafters but have no birdsmouth cuts.

9

Cut the ridge board to size and mark the rafter layout following the SIDE FRAMING illustration (page 117), and then screw the rafters to the ridge. Cut five 1 × 4 collar ties, mitering the ends at 31°. Fasten the collar ties across each set of rafters so the ends of the ties are flush with the rafter edges. Fasten the 2 × 4 crosspiece above the door to the two end rafters. Install remaining cross-pieces as in the FRONT FRAMING (page 116).

10

Install the 1 × 8 siding boards so they overlap the floor frame by 1" at the bottom and extend to the tops of the side walls, and to the tops of the rafters on the front and rear walls. Gap the boards ½", and fasten them to the framing with pairs of 8d galvanized casing nails or siding nails. Install the four 2 × 4 spacers on top of the siding at the front and rear so they match the rafter placement.

11

Cut the arched sections of door trim from 1 × 8 lumber, following the DOOR ARCH TEMPLATE (page 115). Install the arched pieces and straight 1 × 2 side pieces flush with the inside of the door opening. Wrap the window openings with ripped 1 × 6 boards, and then frame the outsides of the openings with 1 × 2 trim. Install a 1 × 2 batten over each siding joint as shown in step 10.

12

Build the 1 × 2 window frames to fit snugly inside the trimmed openings. Assemble the parts with exterior wood glue and galvanized finish nails. If desired, cut a ¼" rabbet in the rear side and install plastic windowpanes with silicone caulk. Secure the window frames in the openings with ⅜" quarter-round molding. Construct the window boxes as shown in the WINDOW BOX DETAIL (page 118). Install the boxes below the windows with 1¼" screws. *(continued)*

To build the spire, start by drawing a line around a 4 × 4 post, 9" from one end. Draw cutting lines to taper each side down to ¾", as shown in SPIRE DETAIL (page 114). Taper the end with a circular saw or handsaw, and then cut off the point at the 9" mark. Cut the post at 43". Add 1 × 2 trim and cap molding as shown in the detail, mitering the ends at the corners. Drill centered pilot holes into the post, balls, and point, and join the parts with dowel screws.

To cut the verge boards, enlarge the VERGE BOARD TEMPLATE (page 114) on a photocopier so the squares measure 1". Draw the pattern on a 1 × 6. Cut the board with a jigsaw. Test-fit the board and adjust as needed. Use the cut board as a pattern to mark and cut the remaining verge boards. Install the boards over the front and rear fascia, then add picture molding along the top edges.

Add a 1 × 2 block under the front end of the ridge board. Center the spire at the roof peak, drill pilot holes, and anchor the post with 6" lag screws. Cut and install the 1 × 6 front fascia to run from the spire to the rafter ends, keeping the fascia ½" above the tops of the rafters. Install the rear fascia so it covers the ridge board. Cut and install two 1 × 4 brackets to fit between the spire post and front fascia, as shown in SPIRE DETAIL (page 114).

Cut the 1 × 6 eave fascia to fit between the verge boards, and install it so it will be flush with the top of the roof sheathing. Cut and install the roof sheathing. Add building paper, metal drip edge, and asphalt shingles.

Mark the deck post locations 1¼" in from the ends and front edge of the porch decking, as shown in the FLOOR PLAN (page 114). Cut four 4 × 4 railing posts at 30". Bevel the top edges of the posts at 45°, as shown in the DECK RAILING DETAIL (page 114). Fasten the posts to the decking and floor frame with 3½" screws. Cut six 2 × 4 treated blocks at 3½". Fasten these to the bottoms of the posts, on the sides that will receive the railings.

Assemble the railing sections following the DECK RAILING DETAIL (page 114). Each section has a 2 × 4 top and bottom rail, two 1 × 2 nailers, and 2 × 2 balusters spaced so the edges of the balusters are no more than 4" apart. You can build the sections completely and then fasten them to the posts and front wall, or you can construct them in place starting with the rails. Cut the shaped trim boards from 1 × 4 lumber, using a jigsaw. Notch the rails to fit around the house battens as needed.

Construct the door with 1 × 6 boards fastened to 2 × 4 Z-bracing, as shown in the DOOR DETAIL (page 115). Fasten the boards to the bracing with glue and 6d finish nails. Cut the square notches and the top of the door with a jigsaw. Add the 2 × 2 brace as shown. Install the door with two hinges, leaving a ¼" gap all around. Add a knob or latch as desired.

Make the foundation "stones" by cutting 116 6"-lengths of 5/4 × 6 deck boards (the pieces in the top row must be ripped down 1"). Round over the cut edges of all pieces with a router. Attach the top row of stones using construction adhesive and 6d galvanized finish nails. Install the bottom row, starting with a half-piece to create a staggered joint pattern. If desired, finish the playhouse interior with plywood or tongue-and-groove siding.

Salt Box Storage Shed

The asymmetrical roofline of this shed, taller in front than in back, resembles the salt box house, a classic of colonial American architecture. Here is a great basic shed that offers lots of storage space with a slightly lower elevation than the Simple Storage Shed. At 77 inches, the doors are still plenty tall enough for almost everyone to enter easily, but the shallow peaked roof and 66-inch rear wall means the overall building height is just under eight feet. The gravel base and 4 × 4 joists (no skids needed) also keep this building close to the ground. These factors all create a smaller visual impact and a pleasing proportional look.

The 8 × 12-foot floor plan and the 6-foot-wide double doors give it excellent versatility and access. The sturdy ¾-inch floor will support a riding mower or other heavy equipment. The centered doors create two feet of space around all sides of parked equipment; however, you could position the doors off center to suit your storage needs.

The construction is quite simple stick framing. Additional features include enclosed soffits to keep out bugs and birds, and a roof vent to prevent overheating. Clapboard style siding would enhance the colonial look of this shed, if you wanted to move away from the standard vertically grooved plywood siding.

The width of this shed can be easily adjusted from 12 feet down to 8 feet or up to 16 feet. The top and bottom plates for the front and rear walls and the ridge board would be shorter or longer, and you would need to subtract or add more studs, joists, and rafters, but the height and depth measurements remain the same.

Examine the plans carefully before you start, as the joists and front and rear wall studs have different spacing on the left side versus the right.

The salt box storage shed draws on classic American architecture to create a space that's as simple on the inside as it is beautiful on the outside.

SHOPPING LIST

DESCRIPTION	QTY/SIZE	MATERIAL
FOUNDATION/FLOOR		
Drainage material	1.25 cu. yd.	Compactible gravel
Skids	7 @ 8'	4 × 4 pressure treated
Rim joists	2 @ 12'	2 × 4 pressure treated
Floor sheathing	3 sheets @ 4 × 8'	¾" exterior-grade plywood
WALL FRAMING		
Bottom plate, front & rear	2 @ 12'	2 × 4
Bottom plate, sides	2 @ 8'	2 × 4
Top plates, front & rear	2 @ 12'	2 × 4
Top plates, sides	2 @ 8'	2 × 4
Studs, wall (cut in two pieces)	10 @ 12'	2 × 4
Studs, front	10 @ 7'	2 × 4
Studs, gable	2 @ 12'	2 × 4
Door header	1 @ 14'	2 × 6
ROOF FRAMING		
Rafters	7 @ 10'	2 × 4
Ridge board	1 @ 12'	2 × 6
Rafter tie	1 @ 12'	2 × 4
EXTERIOR FINISHES		
Siding	11 sheets @ 4 × 8'	⅝" plywood siding
Eave fascia	2 @ 14'	1 × 6 cedar
Gable fascia	2 @ 10'	1 × 4 cedar
Eave soffit	2 @ 12'	1 × 4 cedar
Frieze, rear	1 @ 12'	1 × 4 cedar
Frieze, front ripped	1 @ 12'	1 × 6 cedar
Wall corner trim	4 @ 14'	1 × 4 cedar

DESCRIPTION	QTY/SIZE	MATERIAL
ROOFING		
Sheathing	4 sheets @ 4 × 8	½" exterior-grade plywood roof sheathing
Drip edge	50 linear ft.	Metal drip edge
15# building paper	1 roll	
Shingles	1⅔ squares	Asphalt shingles
DOOR		
Door stop	3 @ 8'	1 × 2 cedar
Door trim & casing	10 @ 8'	1 × 4 cedar
FASTENERS & HARDWARE		
16d galvanized common nails	1 lb.	
16d coated common nails	4 lbs.	
8d coated plywood nails	3 lbs.	
8d galvanized siding nails	5 lbs.	
1¼" galvanized roofing nails	3 lbs.	
8d galvanized casing nails	2 lbs.	
Hinges	6	6" T-strap
3d galvanized door nails	1 lb.	
Exterior caulk	1 tube	
Glue	1 tube	Exterior wood glue
Roof vent	1	

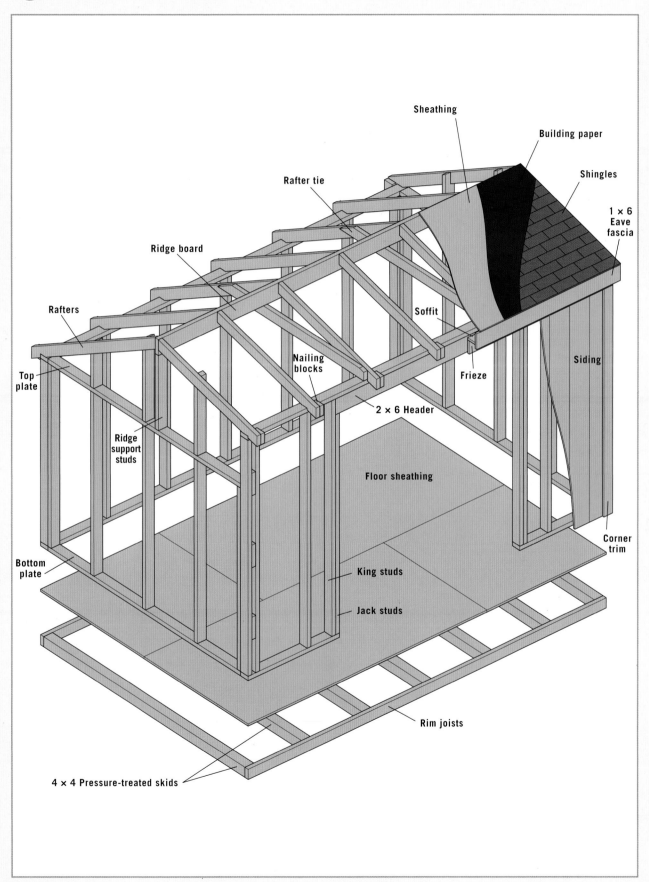

Sheathing

Building paper

Shingles

Rafter tie

1 × 6 Eave fascia

Ridge board

Rafters

Soffit

Top plate

Nailing blocks

Frieze

Siding

Ridge support studs

2 × 6 Header

Floor sheathing

Corner trim

Bottom plate

King studs

Jack studs

Rim joists

4 × 4 Pressure-treated skids

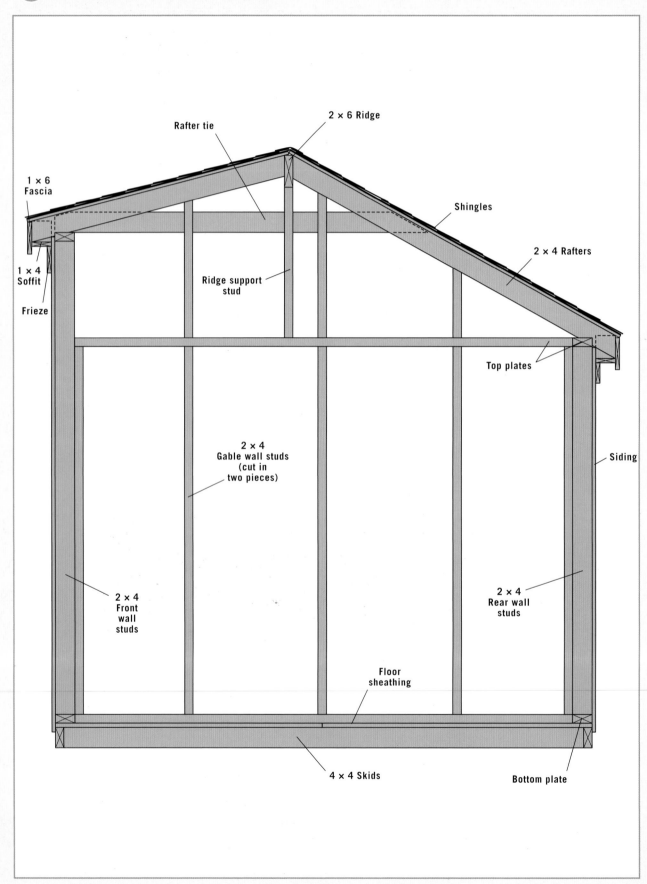

2 × 6 Ridge

Rafter tie

1 × 6
Fascia

Shingles

2 × 4 Rafters

1 × 4
Soffit

Frieze

Ridge support
stud

Top plates

2 × 4
Gable wall studs
(cut in
two pieces)

Siding

2 × 4
Front
wall
studs

2 × 4
Rear wall
studs

Floor
sheathing

4 × 4 Skids

Bottom plate

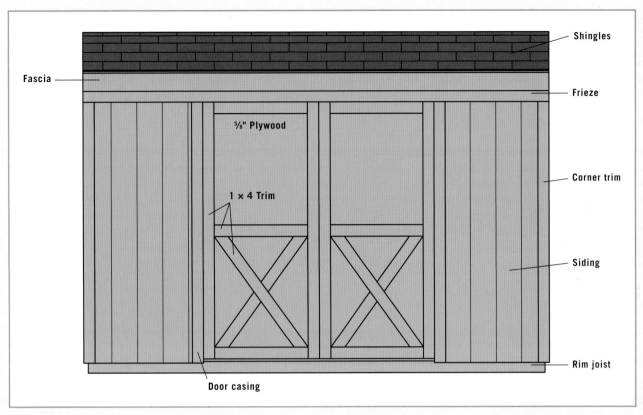

Shingles

Fascia

Frieze

⅝" Plywood

Corner trim

1 × 4 Trim

Siding

Rim joist

Door casing

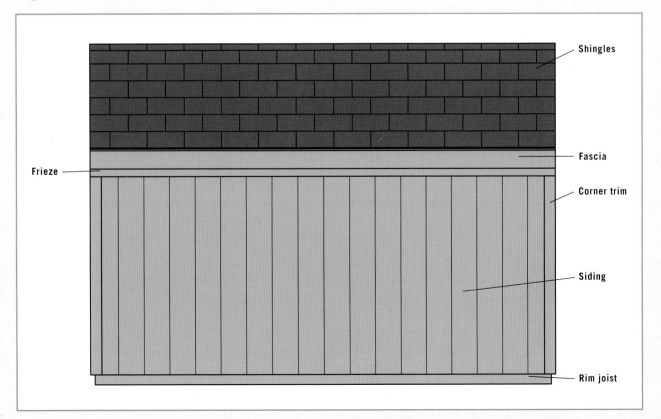

Shingles

Fascia

Frieze

Corner trim

Siding

Rim joist

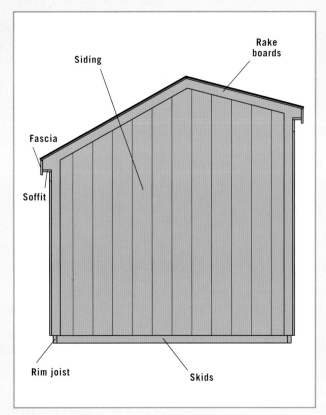

Siding

Rake boards

Fascia

Soffit

Rim joist

Skids

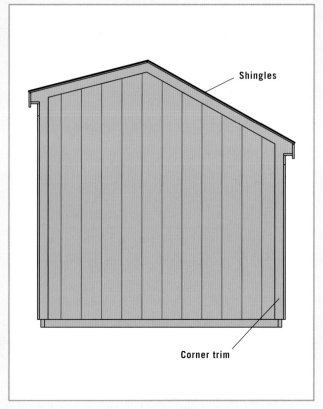

Shingles

Corner trim

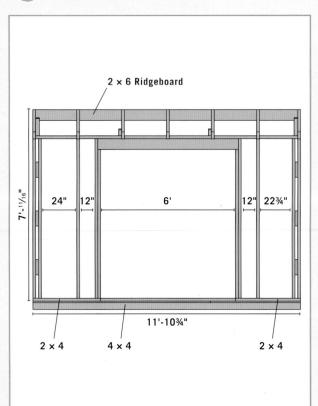

2 × 6 Ridgeboard

7'-1¹¹/₁₆"

24" 12" 6' 12" 22¾"

11'-10¾"

2 × 4   4 × 4   2 × 4

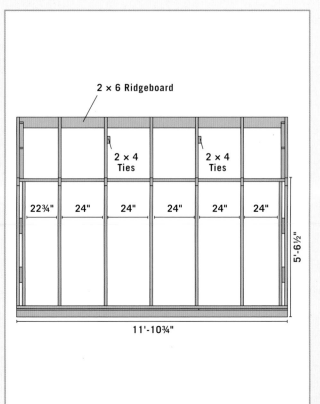

2 × 6 Ridgeboard

2 × 4 Ties   2 × 4 Ties

22¾" 24" 24" 24" 24" 24"

5'-6½"

11'-10¾"

RIGHT FRAMING

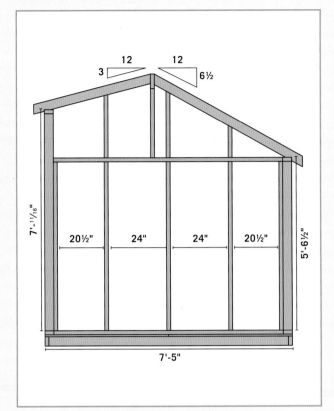

12 12
3 ◁ ▷ 6½

7'-11/16"

20½" 24" 24" 20½"

5'-6½"

7'-5"

LEFT FRAMING

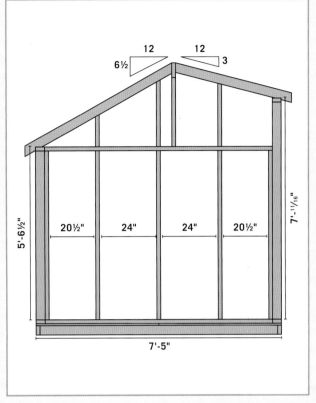

12 12
6½ ◁ ▷ 3

5'-6½"

20½" 24" 24" 20½"

7'-11/16"

7'-5"

FLOOR FRAMING

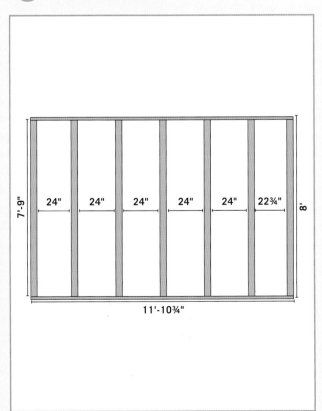

7'-9"

24" 24" 24" 24" 24" 22¾"

8'

11'-10¾"

HEAD DETAIL

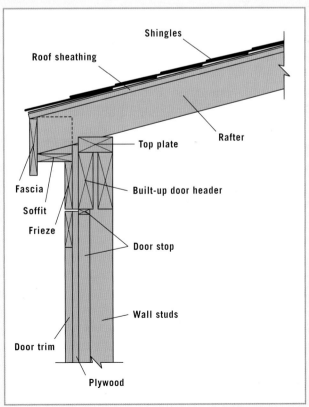

Shingles

Roof sheathing

Top plate

Rafter

Fascia

Built-up door header

Soffit

Frieze

Door stop

Wall studs

Door trim

Plywood

RAKE DETAIL

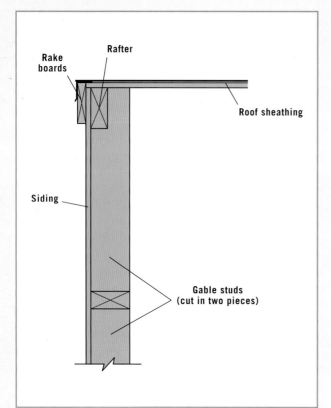

Rake boards

Rafter

Roof sheathing

Siding

Gable studs
(cut in two pieces)

DOOR DETAIL

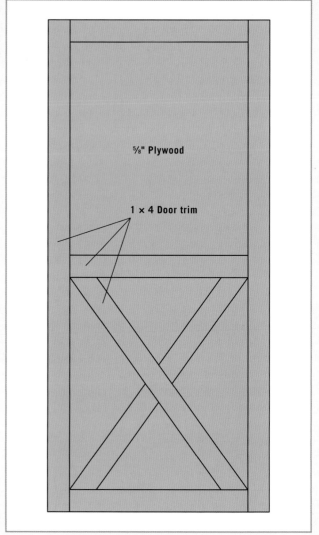

⅝" Plywood

1 × 4 Door trim

CORNER DETAIL

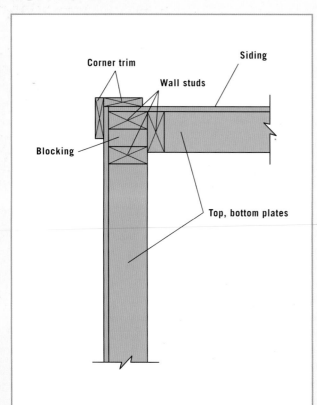

Corner trim

Wall studs

Siding

Blocking

Top, bottom plates

DOOR JAMB DETAIL

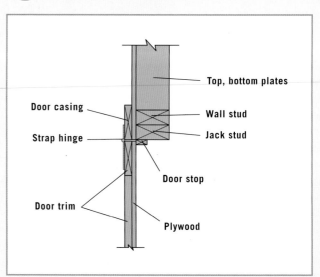

Top, bottom plates

Door casing

Wall stud

Strap hinge

Jack stud

Door stop

Door trim

Plywood

How to Build the Salt Box Shed

Prepare the foundation by following the steps for a simple skid foundation. First, prepare a bed of compacted gravel. Make sure the bed is flat and level. Cut seven 4 × 4" × 8 ft. pressure-treated posts to 93" to serve as skids. Position the joists as shown in the FLOOR FRAMING PLAN on page 131. Note that the right end of the shed is spaced at 21" on center, not 24" on center, as is the rest of the shed. Cut two 12-ft. pressure-treated 2 × 4s to 142¾" for rim joists and fasten to the ends of the 4 × 4 joists using 16d galvanized nails. Measure the diagonals of the foundation to make sure the frame is square.

Install the floor sheathing onto the floor frame. Begin at the left front end of the shed with a full 4 × 8 ft. sheet of ¾" plywood. Attach the plywood with 8d plywood nails, 6" on center around the edges and 10" on center in the field. Cut another 4 × 8 ft. sheet of ¾" plywood into two 4-ft. pieces. Attach one piece at the rear left end of the shed. Cut the second full 4 × 8 ft. sheet to fit the rear right end and attach. Cut the half sheet to fit the front right end.

Begin assembling the rear wall panel by cutting two 2 × 4s at 142¾" for the bottom and top plates. Cut nine 2 × 4s at 63½" for the rear wall studs. Assemble the rear wall according to the REAR WALL FRAMING PLAN on page 130. Nail through the top and bottom plates using two 16d nails at the top and the bottom. Note in the CORNER DETAIL on page 132 that the doubled end studs are spaced 1½" apart with three 6" lengths of 2 × 4 blocking.

Begin assembling the front wall panel by cutting two 2 × 4s at 142¾" for the bottom and top plates. Cut eight 2 × 4s at 81¹¹⁄₁₆" for the wall studs. Assemble the front wall according to the FRONT WALL FRAMING plan on page 130.

(continued)

Cut two 2 × 4s at 76³⁄₁₆" for jack studs. Create the built-up door header by sandwiching a piece of 5½ × 75" piece of ½" plywood between two 2 × 6s cut to 75". Apply construction adhesive to both sides of the plywood. Nail together on both sides with 8d coated common nails or screw together with deck screws. Attach the jack studs and header to the king studs and top and bottom plates.

Begin assembling the sidewall panels by cutting four 2 × 4s at 89" for the sidewall top and bottom plates. Cut ten 2 × 4s at 63½" for the sidewall studs. Assemble the side walls according to the RIGHT SIDE FRAMING and LEFT SIDE FRAMING illustration on page 130. Note that the three middle studs are 24" on center, and the two end studs are spaced 20½" on center.

Tilt the rear wall panel into place. Make sure that the narrow stud spacing is at the right end of the shed. Attach the bottom plate to the rim joist with 16d nails. Plumb the wall and brace into place.

Tilt the sidewall panels into place. Attach the bottom plate to the joists with 16d nails. Check for plumb and nail the panels together at the corners.

Tilt the front panel into place. Attach the bottom plate, except for the portion in the doorway, to the rim joist with 16d nails. Check for plumb and attach to the side walls at the corners.

Install the plywood siding on the front and rear walls. Start on the left hand end of the shed. Cut the siding to length so it is flush with the top of the top plate and overlaps the top of the floor joists by ½". The siding should not extend to the ground as it is not rated for ground contact. Make sure all vertical seams fall on the studs. Attach the siding to the studs with 8d galvanized nails 6" on center at edges and 12" on center in the field.

Cut two 2 × 4 ridge support studs at 26". Center the studs 42" back from the outside face of the front wall stud at each end of the shed. Toenail the studs into place.

Create the ridge board by cutting the 2 × 6 ridge board to 142¾". Set it on top of the ridge support studs and toenail into place. You may need to add braces to hold the ridge support studs plumb while you cut and fit the rafters. *(continued)*

Build the rafters as described in the SHOPPING LIST on page 126. Cut one pair of rafters from a 10-ft.-long 2 × 4. Set the rafters in place, making sure that the ridge support studs are plumb. If the rafters fit well, use them as patterns for cutting the remaining six pairs of rafters. Attach the rafters aligned with the spacing of the wall studs. On the third and fifth rafter pairs, attach 2 × 4 ties. Cut the ties to fit so that they rest atop the front top plate and are level. Attach the ties with 16d nails.

Cut 2 × 4 rake support studs to fit in the gable ends, aligned with the sidewall studs to act as siding nailers. Toenail into place. Measure the sidewall elevation at the middle siding nailer and at the corners, allowing for ½" overlap over the joist. Cut plywood siding to size and apply siding to the side walls.

Cut fourteen 4"-long nailing blocks from scrap 2 × 4s. Attach the nailing blocks to the rafters so that they are flush with the bottom point of the rafter as shown on the HEAD DETAIL, page 131.

Attach the plywood sheathing to the rafters using 8d nails 6" on center on the edges and 12" on center in the field. Make sure all seams fall on rafters. Cut to fit and attach the 1 × 4 cedar rake boards on the gable ends.

To install the roof vent, cut a hole one foot down from the peak of the roof in the middle of the rear roof. Apply roofing felt. Cut the felt away from the vent hole and install the vent. Install the metal drip edging and shingle the roof.

Attach the 1 × 6 cedar fascia to the rafter ends using 8d galvanized trim nails. Attach the 1 × 4 cedar soffit to the nailing blocks. Attach the 1 × 4 cedar frieze board under the soffit as shown on the HEAD DETAIL (page 131). Attach the 1 × 4 cedar corner trim and 1 × 4 cedar doorway trim.

Cut out the bottom plate in the door opening using a reciprocating saw or a back saw. Measure the door opening and adjust door dimensions if necessary. Cut two doors from the plywood siding at 36 × 77½". Attach 1 × 4 cedar trim around the edges of the plywood door blanks with exterior wood glue and 1¼" deck screws. Attach screws from the front and rear. Cut and install the/a decorative X to the bottom, or top and bottom, of each door panel. Install door casing.

Attach three exterior strap hinges to each door, taking care to make them square to the door edge. With a helper, place a door in the opening, flush with the door trim. Use shims to create a ¼" gap at the side and top. Attach the hinges. Repeat for the second door. Install 1 × 2 cedar door stops around the inside of the door frame. Attach a cedar 1 × 4 on the reverse of the left door with a 2" overhang to act as a stop. Attach desired handles and hasp to the doors. Paint or seal the shed.

Solar Power for Your Shed

Electrical power can be a wonderful addition to a mid-size or larger shed. It can allow you to use the shed at night or on cloudy days, and make it possible to charge electric yard equipment, small power tools, and even a computer for a shed-based home office. But running a power line from your home to a shed usually involves a great deal of effort and expense.

A roof-mounted solar photovoltaic panel can offer a cutting-edge, environmentally friendly alternative. Although you can cobble together the necessary parts from different manufacturers or retailers, the easiest way to wire a shed for solar power is to buy an all-in-one kit.

Many solar shed kits are meant specifically to power a light fixture, to add safety and security to the shed, and to allow for nighttime use. However, kit manufacturers also supply systems that power both lights and outlets, so that you have more options. These kits typically come equipped with a photovoltaic panel and mounting hardware, a battery to store the electrical charge when the system isn't being used or at night, an inverter (a device that converts the DC power coming from the solar panel into more usable AC power), a pass-thru box, a power box, and a lighting fixture or fixtures. You can find these systems at specialty suppliers or widely available online (see Resources, page 250).

Keep in mind that not all sheds are good candidates for solar power. First, determine if your shed's exposure allows for a reasonable return on investment (if you haven't yet built your shed, you can site it with optimal exposure in mind). It's all about how much direct sunlight the roof receives. There are a number of ways to check a shed's solar exposure. You can use a tool such as a Solar Pathfinder, although these are rarely available through tool rental stores and are usually too expensive for one-time use. Or, you may be able to use search engine map features—many include a street address lookup feature that allows you to check how much direct sun your property receives.

In any case, strive for an exposure that is as close to due south as possible. The shed's roof should not be shaded by nearby trees or structures, and the roof design itself must allow for placement of the panels at an angle to the southern exposure (on flat roofs, you can mount a panel with feet that are designed to raise one end).

Installing a solar panel requires only a basic understanding of electrical systems, and modest DIY skills. However, if you don't feel comfortable working with an electrical system, hire an electrician. A reputable electrician will be able to install the system in less than a half day, so the added expense should not be prohibitive. The steps that follow will give you a good idea of what's involved in the installation.

Manufacturer's supply turnkey solar power kits specifically meant for powering sheds. The most common of these supply everything you'll need to install the system, except for the tools.

A solar panel is a small addition to your shed, but one that can pay big dividends in increasing the usability of the structure. Simple to install, the panels are compact enough that they don't detract from the shed's look.

Most solar power kits for sheds include at least one light, such as the LED fixture shown here. It's often a good idea to opt for a kit with exterior lights as well, to add safety and security to the structure—especially if you plan on storing valuable power equipment and tools in the shed. Add a motion detector to the light fixture for even more security. (The power box is the white cube mounted on the wall right below the light.)

As with any electrical system, use extreme care when working with a solar panel system. Do not try to install the system if you have any doubt about your abilities or do not entirely understand the process. If you decide to perform the installation yourself, pay attention to the following points:

- Keep in mind that even if a breaker or switch is off, a solar panel is energized anytime there is sunlight.

- Don't work on the system in wet or icy weather.

- Use insulated tools when working with the battery, and don't wear jewelry when working with a solar power system.

- Solar batteries can be very heavy, so use a helper as necessary to move or place the battery.

- It's also smart to use a helper when installing the panels. They can be unwieldy, and working on a pitched roof can mean you're off balance.

The mounting stand for the PV panel is constructed from metal U-channel and pre-bent fasteners (a product called Unistrut is seen here; see Resources, page 250). Position the solar panel where it will receive the greatest amount of sunlight for the longest period of time each day—typically the south-facing side of a roof or wall. For a circuit with a battery reserve that powers two to four 12-volt lights, a collection panel rated between 40 and 80 watts of output should suffice. These panels can range from $200 to $600 in price, depending on the output and the overall quality.

Schematic Diagram for an Off-the-Grid Solar Lighting System

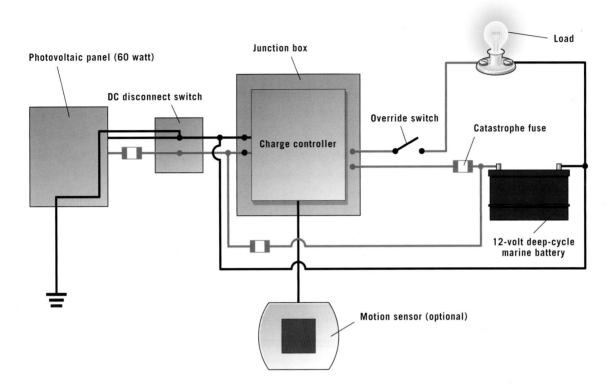

- Photovoltaic panel (60 watt)
- DC disconnect switch
- Junction box
- Charge controller
- Override switch
- Load
- Catastrophe fuse
- 12-volt deep-cycle marine battery
- Motion sensor (optional)

TOOLS & MATERIALS

- Tape measure
- Drill/driver with bits
- Caulk gun
- Wiring tools
- Metal-cutting saw
- Socket wrench
- Photovoltaic panel (40 to 80 watts)
- Charge controller
- Catastrophe fuse
- Battery sized for 3-day autonomy
- Battery case
- Battery cables
- 12-volt LED lights, including motion-sensor light
- Additional 12-volt light fixtures as desired
- 20 ft. Unistrut 1⅞"-thick U-channel (see Resources, page 250)

- 45° Unistrut connectors
- 90° Unistrut angle brackets
- Unistrut hold-down clamps
- ⅜" spring nuts
- ⅜"-dia. × 1"-long hex-head bolts with washers
- Green ground screws
- DC-rated disconnect or double throw snap switch
- 6" length of ½"-dia. liquid-tight flexible metallic conduit
- ½" liquid-tight connectors
- Lay-in grounding lugs
- Insulated terminal bars to accept one 2-gauge wire and four 12-gauge wires
- Cord cap connectors for ½"-dia. cable
- ½" ground rod & clamp

- Copper wire (6- & 14-gauge)
- Square boxes with covers
- ½" flexible metallic conduit or Greenfield
- ½" Greenfield connectors
- 1¹⁄₁₆" junction boxes with covers
- PVC 6 × 6" junction box with cover
- 14/2 UF wire
- ¼" × 20 nuts & bolts with lock washers
- Roof flashing boot
- Roof cement
- Silicone caulk
- Eye protection

Check the underside of the roof for electrical lines or ductwork. Locate an access hole for the panel cable into the roof deck, directly behind the planned location for the panel assembly. Buy a flashing boot with a rubber boot sized for small electrical conduit (¾" or 1" is best). Place the boot so that the top edge extends under two shingle tabs, then drill a test hole with a ¼" bit. Leave the bit in place and double-check the underside of the roof to make sure you come out in the right spot. If everything looks good, finish the hole with a hole saw or spade bit big enough for the conduit to fit through.

Cut a piece of conduit long enough to go through the roof and extend several inches above the boot. The conduit should continue on the underside of the roof over to the planned location of the regulator. Push the pipe through and hold it in place with a pipe strap or block of wood. Slip the roof boot over the pipe and wiggle it into place under the shingles. Spread roofing cement under the sides (but not the bottom) of the metal flashing, then nail it to the roof. Glue a 90° elbow to the conduit, turning it downhill, and extend with more conduit (if necessary) to the location for the PV assembly. Fish the leads from the panels through the conduit to the regulator. After running the wires, plug the opening in the conduit around the wires with electrician's putty or caulk to seal out bugs and drafts.

 ## How to Hook Up a Solar Module

(1)

(2)

Install the mount for the solar module on the roof. The stand components shown here are held together with bolts and spring-loaded fasteners. The 45° and 90° connectors are manufactured specifically for use with this Unistrut system.

Secure the module to the mount and run feed wires from the module through the roof penetration (see above) and into the shed. Connections for the feed wires that carry current from the collector are made inside an electrical box mounted on the back of the collector panel.

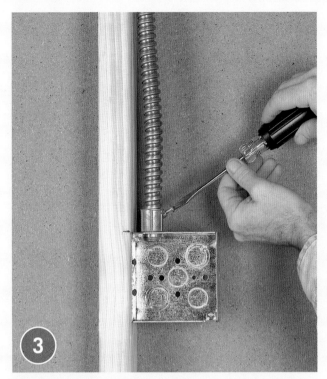

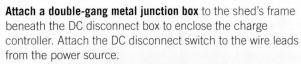

Attach a double-gang metal junction box to the shed's frame beneath the DC disconnect box to enclose the charge controller. Attach the DC disconnect switch to the wire leads from the power source.

Mount a junction box inside the building where the conduit and wiring enter from the power source. Secure the box to the conduit with appropriate connectors. The two 14-gauge wires running through the conduit are connected to the positive and negative terminals on the panel on the roof side of the shed. Run flexible metal conduit from the entry point at the power source to the junction box for the DC disconnect box. Use hangers rated for flexible conduit.

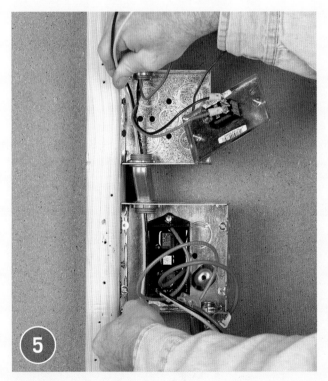

Install the charge controller inside the double-gang box. Run flexible conduit with connectors and conductors from the disconnect box and to the charge controller box.

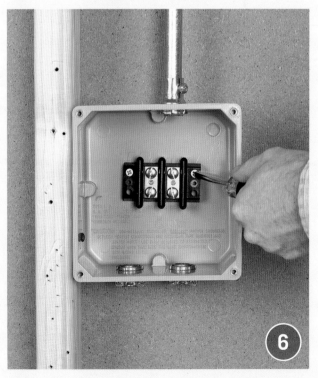

Mount a PVC junction box for the battery controller about 2 ft. above the battery location and install two insulated terminal bars within the box. *(continued)*

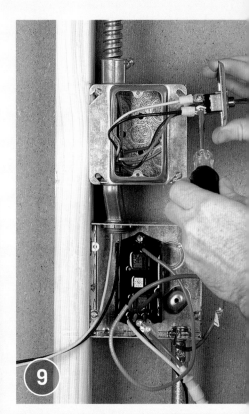

Set up grounding protection. Pound an 8-ft.-long, ½"-dia. ground rod into the ground outside the building, about 1 ft. from the wall on the opposite side of the charge controller. Leave about 2" of the rod sticking out of the ground. Attach a ground rod clamp to the top of the rod. Drill a ⁵⁄₁₆" hole through the garage wall (underneath a shake or siding piece) and run the #6-gauge THWN wire to the ground rod. This ground will facilitate lightning protection.

Build a support shelf for the battery using 2 × 4s. The shelf should be at least 18" above ground. Set the battery on the shelf in a sturdy plastic case.

Wire the DC disconnect. Attach the two #14-gauge wires to the two terminals labeled "line" on the top of the DC disconnect switch.

Fuse

Run wiring to the loads (light fixtures) from the charge controller. DC light fixtures (12-volt) with LED bulbs can be purchased at marine and RV stores if you can't find them in your home center or electrical supply store. If you prefer, you can wire an AC inverter into your solar system to provide alternating current to standard lights or even a receptacle.

Wire the charge controller. Route two more #14-gauge wires from the bottom of the DC disconnect terminals into the 4 × ¹¹⁄₁₆" junction box and connect to the "Solar Panel In" terminals on the charge controller. The black wire should connect to the negative terminal in the PVC box and the red to the positive lead on the charge controller. Finish wiring of the charge controller according to the line diagram provided with the type of controller purchased. Generally the load wires connect to the orange lead and the red wire gets tied to the battery through a fuse.

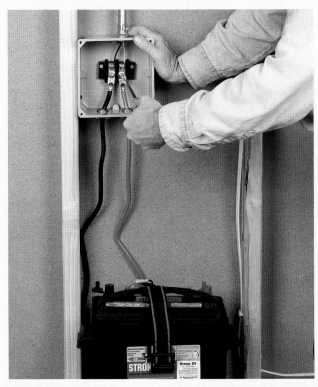

Install the battery. Here, a deep-cycle 12-volt marine battery is used. First, cut and strip each of the two battery cables at one end and install into the battery control junction box through cord cap connectors. Terminate these wires on two separate, firmly mounted insulated terminal blocks.

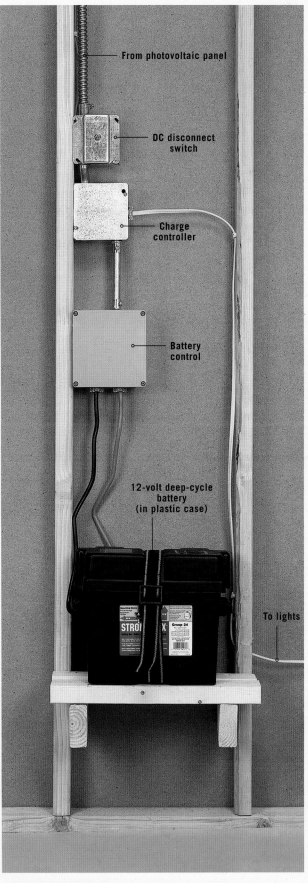

From photovoltaic panel

DC disconnect switch

Charge controller

Battery control

12-volt deep-cycle battery (in plastic case)

To lights

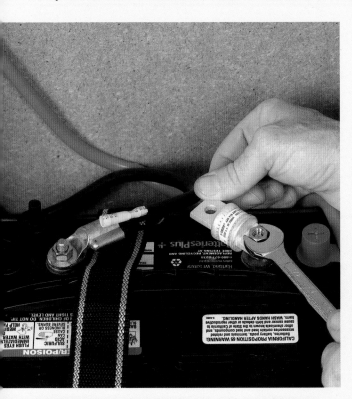

Install the catastrophe fuse onto the positive terminal using nuts and bolts provided with the battery cables. Connect the battery cables to the battery while paying close attention to the polarity (red to positive and black to negative). Make sure all connections have been made and double checked.

Cover all junction boxes, then remove the bag from the panel and turn the DC disconnect switch on to complete the circuit. Test the lights and adjust the time to desired setting.

Large Sheds

Large sheds are usually more complex structures. This means that in addition to requiring more materials and labor, it also involves more details and variables. The reward for your extra work is something closer to a room addition than a simple backyard storage structure.

Larger sheds open up all kinds of possible uses. Many are modified to serve as garages (for cars, motorcycles, and even recreational vehicles such as snowmobiles), complete with service bay, workbenches, and extra room to maneuver. But large sheds—especially larger prefab models—are increasingly being used to supplement living space. A large shed can serve as a simple guest cottage, a band practice area, or just an outdoor dining space with indoor amenities.

Whatever your intended use for a large shed, keep in mind that planning becomes even more important than on smaller structures. Specifically, codes will determine just how large (and tall) your large shed can be and the services that you can reasonably run to the shed. That's important, because you often get the most use out of larger sheds if they have a source of power and even a water supply.

Create the ideal outdoor room with the right large shed at the edge of a yard. Here, a simple modern structure is fitted with accordion doors for a warm-weather outdoor room, and is paired with a dining deck that makes it even more inviting and attractive.

Build a home for your car with the right shed. This stunning, modern structure combines a handy detached garage that could be placed exactly where the homeowner wanted it, with enough room for a light-filled workshop area. Paint and style the shed to match your home to make it look like a natural part of the property that has been there forever.

Install a cheap, three-season oasis with a shed instead of building a home extension. This throwback summer cottage may not have power or plumbing, but it does have comfort and room to spare. The windows can be replaced with screens in the summer, and when furnished accordingly it can actually serve as a lovely privacy guesthouse in the warmer months, all for far less than a home addition would cost.

Opt for modern to really make a design statement. A sleek shed such as this is an incredible visual inside and out. This pre-fab model makes bringing high style to a corner of your yard a simple process.

Make a style statement with a pool shed. This pool house is an example of how a shed design can serve as a stunning showpiece. In addition to the impressive appearance, the pool house provides abundant room for meals or just relaxing out of the hot sun.

Kick your garage up a notch by building a car shed that looks like much more. With its faux-barn-door garage door and natural wood trim, this detached structure is more stylish than most car shelters. The peaked roof and gable allow for a spacious loft inside.

Accent a large, impressive shed design to draw more attention to the structure and make the most of your money and effort. This modern pool cabana was already remarkable, with sliding glass doors that take in an impressive view and a sleek modern architecture that perfectly complements a high-end lap pool. But the natural wood screens with boards running in two different directions emphasize the sophistication of this large shed.

A Quaker-style gable shed is an evocative symbol of heartland America, and the ample dimensions of this design make this a bridge between shed and barn. As nice as the simple appearance is, the abundant open storage inside means just about anything can be kept out of sight, ensuring a tidy yard.

Blur the line between shed and cottage by detailing and outfitting a guesthouse shed to mimic the main house. Here, a shed has been built at a landing along stairs that run from the driveway to the house. The shed is painted and detailed just like the house. With power and water, it is a perfect guest residence that is even cozier and more private than a guest bedroom.

Make outdoor entertaining stylish with some over-the-top features added to your shed. This wonderful yard addition features elegant French doors, a café pass-through window with bar, and even exterior speakers for the sound system inside. This is one shed that can serve as the life of the party.

Serve many functions with a large, woodsy hideaway. Clad in appealing stained wood, and fit with a wealth of windows and a bank of skylights, this stunning shed offers plenty of room for relaxing "away from it all" in your own backyard, or for potting next season's crops. Whatever you do in a structure like this, you'll have plenty of room—and peaceful surroundings—to do it in.

Complement a large shed with a large sitting area. If you're going to use your large shed for relaxing, dining, or any pure enjoyment activity, it pays to add a sitting area outside. Here, a rustic shed is paired with a tiered deck bordering a pond—the stuff of which daydreams are made.

Capitalize on the nature of sheds by making the ultimate custom garage. Building a shed for your car or motorcycle is a way to keep the vehicle stored out of the elements while adding a focal point to the house and yard. The modern garage (top) drips in style with a raked roof and frosted glass inserts in the door. The stately example (left) shines with a stained hardwood door topped by divided lites, and side windows that add illumination to the interior space and a bit of lovely style to the exterior.

Clerestory Studio

This easy-to-build shed is made distinctive by its three clerestory windows on the front. In addition to their unique architectural effect, clerestory windows offer some practical advantages over standard windows. First, their position at the top of the building allows sunlight to spread downward over the interior space to maximize illumination. Most of the light is indirect, creating a soft glow without the harsh glare of direct sunlight. Clerestories also save on wall space and offer more privacy and security than windows at eye level. These characteristics make this shed design a great choice for a backyard office, artist's studio, or even a remote spot for the musically inclined to get together and jam.

As shown, the clerestory studio has a 10 × 10-foot floorplan. It can be outfitted with double doors that open up to a 5-foot-wide opening, as seen here. But if you don't need a door that large, you can pick up about 2½ feet of additional (and highly prized) wall space by framing the opening for a 30-inch-wide door. The studio's striking roofline is created by two shed-style roof planes, which makes for deceptively easy construction.

The shed's walls and floor follow standard stick-frame construction. For simplicity, you can frame the square portions of the lower walls first, then piece in the framing for the four "rake," or angled, wall sections. To support the roof rafters, the clerestory wall has two large headers (beams) that run the full length of the building. These and the door header are all made with standard 2× lumber and a ½-inch plywood spacer.

You can increase the natural light in your studio—and add some passive solar heating—by including the two optional skylights. To prevent leaks, be sure to carefully seal around the glazing and the skylight frame. Flashing around the frame will provide an extra measure of protection.

The clerestory windows on this sunlight-filled shed introduce natural light without introducing a security risk.

SHOPPING LIST

DESCRIPTION	QTY/SIZE	MATERIAL
FOUNDATION		
Drainage material	1.5 cu. yd.	Compactible gravel
Skids	2 @ 10'	4 × 6 pressure-treated landscape timbers
FLOOR		
Rim joists	2 @ 10'	2 × 6 pressure treated
Joists	9 @ 10'	2 × 6 pressure treated
Floor sheathing	4 sheets, 4 × 8'	¾" tongue-&-groove ext.-grade plywood
WALL FRAMING		
Bottom plates	4 @ 10'	2 × 4
Top plates, front walls	5 @ 10'	2 × 4
Top plates, rear wall	2 @ 10'	2 × 4
Top plates, side walls	6 @ 10'	2 × 4
Studs, rear wall	11 @ 8'	2 × 4
Studs, front wall (& clerestory wall)	11 @ 8'	2 × 4
Studs, side walls	26 @ 8'	2 × 4
Header, above windows	2 @ 10'	2 × 6
Header, below windows	2 @ 10'	2 × 10
Header, door	2 @ 8'	2 × 6
Header & post spacers		See Sheathing, below
ROOF FRAMING		
Rafters (& blocking)	20 @ 8'	2 × 6
EXTERIOR FINISHES		
Sidewall fascia	4 @ 8'	2 × 6
Eave fascia	3 @ 12'	2 × 6
Fascia drip edge	8 @ 8'	1 × 2
Siding	10 sheets @ 4 × 8'	⅝" Texture 1-11 plywood siding
Corner trim	10 @ 8'	1 × 4 cedar
Bottom siding trim	5 @ 12'	1 × 4 cedar
Vents	8	2"-dia. round metal vents
ROOFING		
Sheathing (& header/post spacers)	6 sheets @ 4 × 8'	½" exterior-grade plywood roof sheathing
15# building paper	1 roll	
Shingles	1⅔ squares	Asphalt shingles— 250# per sq. min.
Roof flashing	10'-6"	Aluminum

DESCRIPTION	QTY/SIZE	MATERIAL
WINDOWS		
Glazing	3 pieces @ 21 × 36"	¼"-thick acrylic or polycarbonate glazing
Window stops	5 @ 8'	1 × 2 cedar
Glazing tape	60 linear ft.	
Clear exterior caulk	1 tube	
DOOR		
Panels	2 sheets @ 4 × 8'	¾" exterior-grade plywood
Panel trim	8 @ 8'	1 × 4 cedar
Stops	3 @ 8'	1 × 2 cedar
Flashing	6 linear ft.	Aluminum
SKYLIGHTS (OPTIONAL)		
Glazing	2 pieces @ 13 × 22½"	¼"-thick plastic or polycarbonate glazing
Frame	2 @ 8'	1 × 4 cedar
Stops	2 @ 8'	1 × 2 cedar
Glazing tape	25 linear ft.	
FASTENERS & HARDWARE		
16d galvanized common nails	4 lbs.	
16d common nails	16½ lbs.	
10d common nails	1 lb.	
8d galvanized common nails	3 lbs.	
8d box nails	3½ lbs.	
8d galvanized siding nails	7 lbs.	
1" galvanized roofing nails	5 lbs.	
8d galvanized casing nails	2 lbs.	
1¼" galvanized screws	1 lb.	
2" galvanized screws	1 lb.	
Door hinges with screws	6 @ 3½"	
Door handle	2	
Door lock (optional)	1	

10 × 10 FRONT ELEVATION

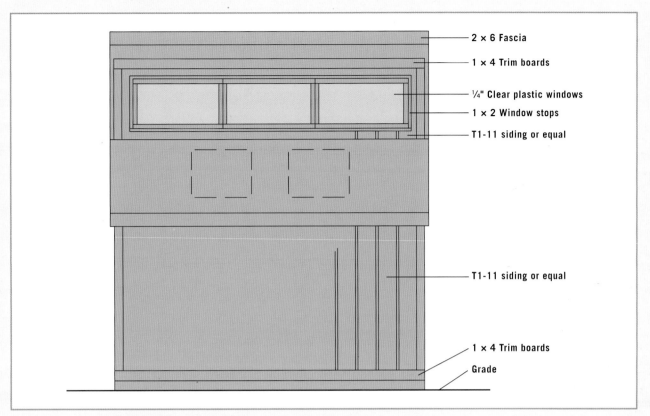

- 2 × 6 Fascia
- 1 × 4 Trim boards
- ¼" Clear plastic windows
- 1 × 2 Window stops
- T1-11 siding or equal
- T1-11 siding or equal
- 1 × 4 Trim boards
- Grade

10 × 10 REAR ELEVATION

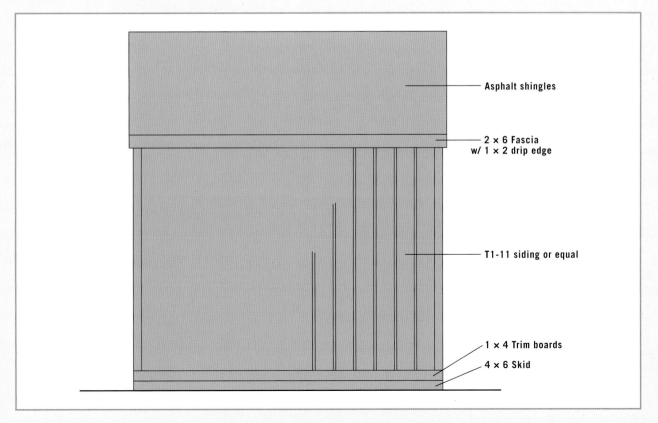

- Asphalt shingles
- 2 × 6 Fascia w/ 1 × 2 drip edge
- T1-11 siding or equal
- 1 × 4 Trim boards
- 4 × 6 Skid

BUILDING SECTION

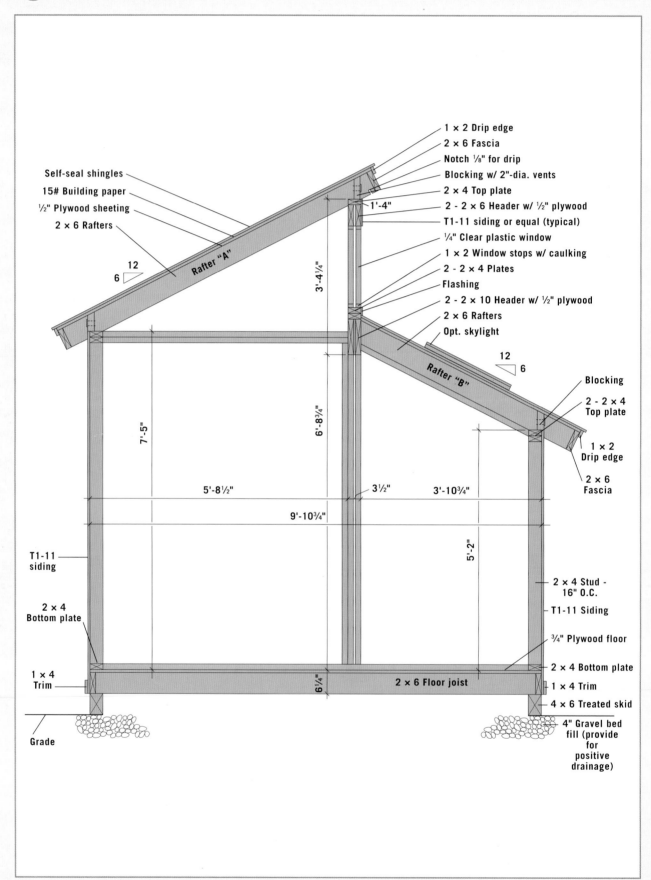

Self-seal shingles
15# Building paper
½" Plywood sheeting
2 × 6 Rafters

Rafter "A"

6 — 12

1 × 2 Drip edge
2 × 6 Fascia
Notch ⅛" for drip
Blocking w/ 2"-dia. vents
2 × 4 Top plate
2 - 2 × 6 Header w/ ½" plywood
T1-11 siding or equal (typical)
¼" Clear plastic window
1 × 2 Window stops w/ caulking
2 - 2 × 4 Plates
Flashing
2 - 2 × 10 Header w/ ½" plywood
2 × 6 Rafters
Opt. skylight

1'-4"

3'-4¼"

Rafter "B"

12 — 6

Blocking
2 - 2 × 4 Top plate
1 × 2 Drip edge
2 × 6 Fascia

6'-8¾"

7'-5"

5'-8½"   3½"   3'-10¾"

9'-10¾"

5'-2"

T1-11 siding

2 × 4 Bottom plate

1 × 4 Trim

Grade

6¼"

2 × 6 Floor joist

2 × 4 Stud - 16" O.C.
T1-11 Siding
¾" Plywood floor
2 × 4 Bottom plate
1 × 4 Trim
4 × 6 Treated skid
4" Gravel bed fill (provide for positive drainage)

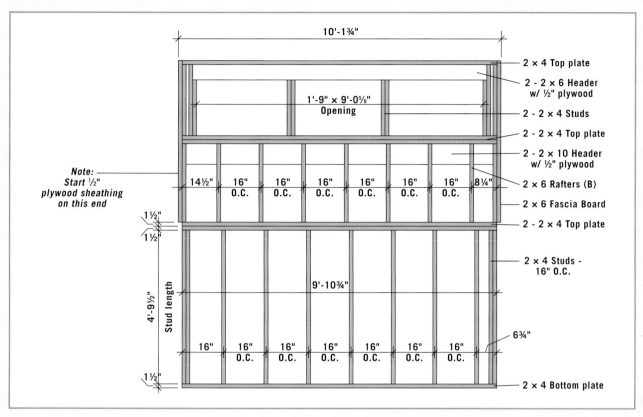

10'-1¾"

2 × 4 Top plate

2 - 2 × 6 Header w/ ½" plywood

1'-9" × 9'-0⅝" Opening

2 - 2 × 4 Studs

2 - 2 × 4 Top plate

2 - 2 × 10 Header w/ ½" plywood

Note: Start ½" plywood sheathing on this end

14½"   16" O.C.   16" O.C.   16" O.C.   16" O.C.   16" O.C.   16" O.C.   8¼"

2 × 6 Rafters (B)

2 × 6 Fascia Board

2 - 2 × 4 Top plate

1½"
1½"

2 × 4 Studs - 16" O.C.

4'-9½" Stud length

9'-10¾"

6¾"

16" O.C.   16" O.C.   16" O.C.   16" O.C.   16" O.C.   16" O.C.   16" O.C.

1½"

2 × 4 Bottom plate

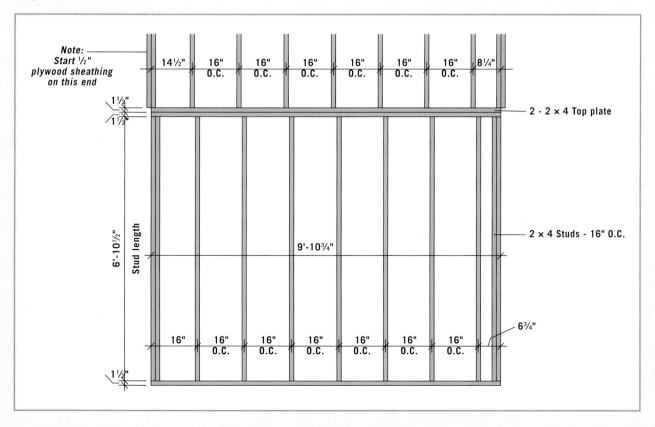

Note: Start ½" plywood sheathing on this end

14½"   16" O.C.   16" O.C.   16" O.C.   16" O.C.   16" O.C.   16" O.C.   8¼"

1½"
1½"

2 - 2 × 4 Top plate

6'-10½" Stud length

9'-10¾"

2 × 4 Studs - 16" O.C.

6¾"

16" O.C.   16" O.C.   16" O.C.   16" O.C.   16" O.C.   16" O.C.   16" O.C.

1½"

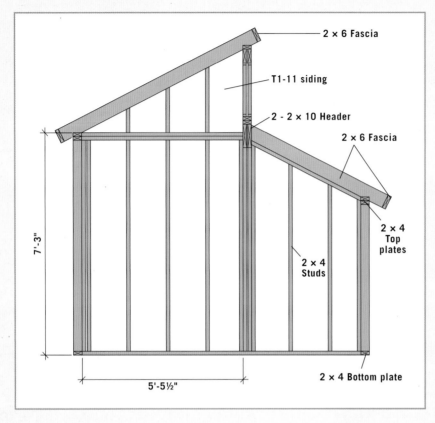

2 × 6 Fascia

T1-11 siding

2 - 2 × 10 Header

2 × 6 Fascia

2 × 4 Top plates

2 × 4 Studs

7'-3"

5'-5½"

2 × 4 Bottom plate

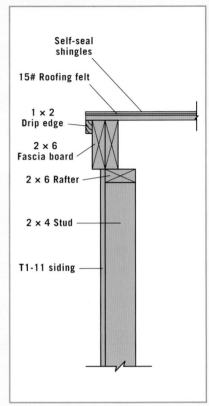

Self-seal shingles

15# Roofing felt

1 × 2 Drip edge

2 × 6 Fascia board

2 × 6 Rafter

2 × 4 Stud

T1-11 siding

DOOR DETAIL

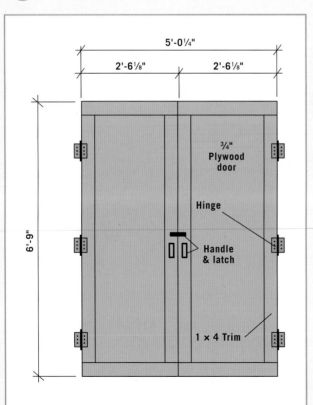

5'-0¼"

2'-6⅛"

2'-6⅛"

¾" Plywood door

Hinge

Handle & latch

6'-9"

1 × 4 Trim

JAMB/CORNER DETAIL

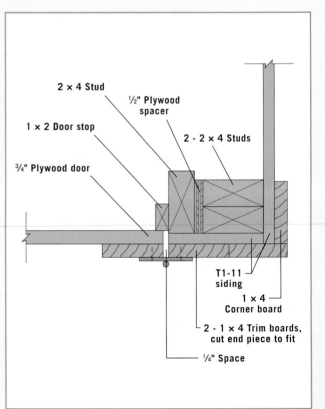

2 × 4 Stud

½" Plywood spacer

1 × 2 Door stop

2 - 2 × 4 Studs

¾" Plywood door

T1-11 siding

1 × 4 Corner board

2 - 1 × 4 Trim boards, cut end piece to fit

¼" Space

RIGHT ELEVATION

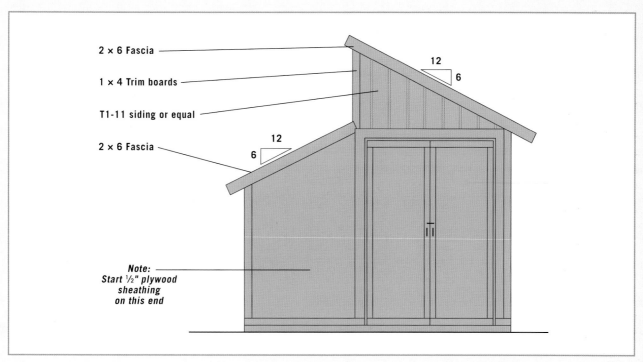

2 × 6 Fascia

1 × 4 Trim boards

T1-11 siding or equal

2 × 6 Fascia

12
6

12
6

Note:
Start ½" plywood
sheathing
on this end

 FLOOR PLAN

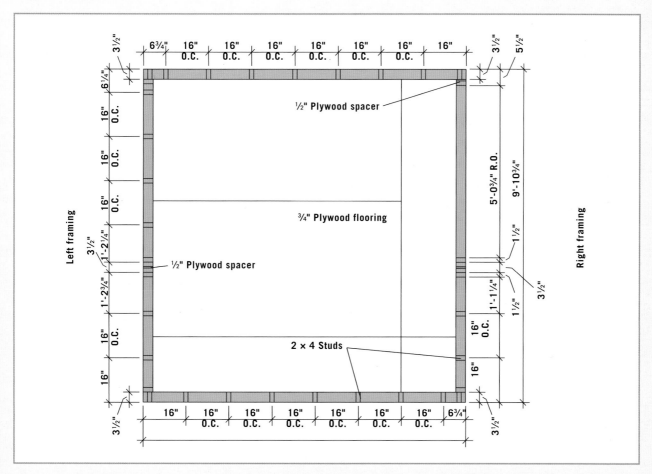

3½"
6¾"   16"   16"   16"   16"   16"   16"   16"   3½"   5½"
O.C.   O.C.   O.C.   O.C.   O.C.   O.C.   O.C.

6¼"

½" Plywood spacer

16"
O.C.

16"
O.C.

5-0¾" R.O.
9'-10¾"

16"
O.C.

¾" Plywood flooring

Left framing

3½"
1'-2¼"

1½"

Right framing

1'-2¾"

½" Plywood spacer

1-1¼"
1½"
3½"

16"
O.C.

16"
O.C.

2 × 4 Studs

16"
O.C.

16"

16"   16"   16"   16"   16"   16"   16"   6¾"
O.C.   O.C.   O.C.   O.C.   O.C.   O.C.   O.C.

3½"
3½"

RAFTER TEMPLATE (A)

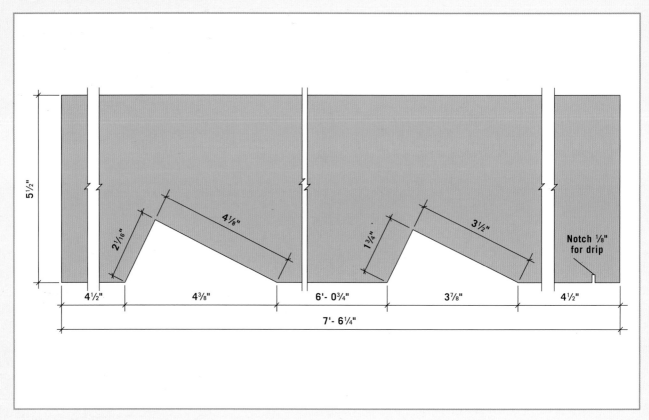

5½"

2¹/₁₆"

4⅛"

1¾"

3½"

Notch ⅛"
for drip

4½"

4⅜"

6'- 0¾"

3⅞"

4½"

7'- 6¼"

RAFTER TEMPLATE (B)

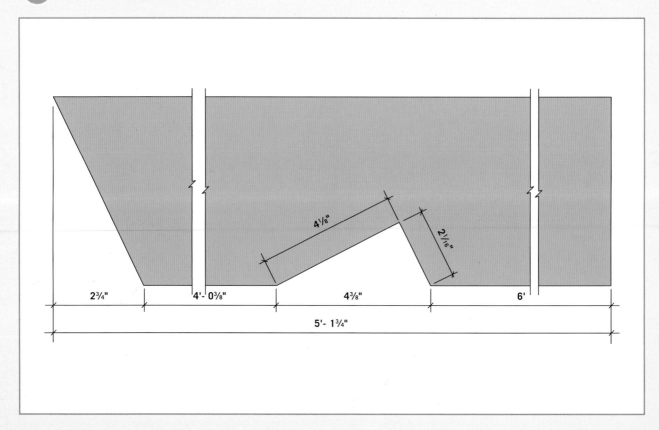

4⅛"

2¹/₁₆"

2¾"

4'- 0⅜"

4⅜"

6'

5'- 1¾"

How to Build the Clerestory Studio

Prepare the foundation site with a 4"-deep layer of compacted gravel. Cut the two 4 × 6 timber skids at 118¾". Position the skids on the gravel bed so their outside edges are 118¾" apart, making sure they are level and parallel with one another.

Cut two 2 × 6 rim joists at 118¾". Cut nine 2 × 6 joists at 115¾". Build the floor frame on the skids and measure the diagonals to make sure the frame is square. Toenail the rim joists to the skids with 16d galvanized common nails.

Install floor sheathing onto the floor frame, starting at the left rear corner of the shed, as shown in the FLOOR PLAN (page 161). Rip the two outer pieces and final corner piece so their outside edges are flush with the sides of foundation skids.

Cut the studs and top and bottom plates for the front wall and nail them together with 8d common nails. Position the wall on the floor deck and raise it. Fasten it by driving 16d common nails through the sole plate and into the floor deck and frame. *(continued)*

Assemble the rear wall framing with a bottom plate and double top plate at 118¾" and 82½" studs 16" on-center. Assemble the square portions of the left and right side walls. Attach the rear wall and nail the side walls in place.

Take measurements to confirm the dimensions for the clerestory wall frame. Build the clerestory frame wall to match the dimensions.

Construct the sloped portions of the side walls. Install them by nailing them to the floor deck with 16d common nails. Also nail the corners to the front wall.

Create the headers by sandwiching a ½" plywood strip between two 2× dimensional framing members. Assemble the header with deck screws driven through both faces.

Set the main header on top of the sidewall posts and toenail it in place with 16d common nails. The main header ends should be flush with the outsides of the side walls.

Lift the clerestory wall frame onto the main header. Orient the wall so it is flush in front and on the ends, and then attach it to the main header with 16d nails.

Install T1-11 siding on the front wall, starting at the left side (when facing the front of the shed). Cut the siding to length so it's flush with the top of the top plate and the bottom of the floor joists. Make sure any vertical seams fall at stud locations. Add strips of siding to cover the framing on the clerestory wall. Install siding on the rear wall, starting at the left side (when facing the rear side of the shed).

Cut one of each "A" and "B" pattern rafters from a single 16-ft. 2 × 6, using the RAFTER TEMPLATES (page 162). Both roof planes have a 6-in-12 slope. Test-fit the rafters and make any necessary adjustments, then use the patterns to cut eight more rafters of each type. Install the rafters as shown in REAR FRAMING and FRONT FRAMING (page 159). Toenail the top ends of the "B" rafters to the main header. *(continued)*

Frame each of the upper rake walls following the same technique used for the gable walls. Cut the top plate to fit between the clerestory header and the door header (on the right side wall) or the top plate (on the left side wall). Install four studs in each wall using 16" on-center spacing.

Install 2 × 6 fascia boards flush with the top edges and ends of the rafters. The upper roof gets fascia on all four sides; the lower roof on three sides. Miter the corner joints if desired. Install siding on the side walls, flush with the bottom of the fascia; see the RAKE DETAIL (page 160.)

Install ½" plywood roof sheathing, starting at the bottom left side of the roof on both sides of the shed. Run the sheathing to the outside edge of the fascia. Add 1 × 2 trim to serve as a drip edge along all fascia boards, flush with the top of the sheathing.

Fasten 1 × 2 stops inside the window rough openings, flush with the inside edges of the framing, using 2" screws. Set each window panel into its opening, using glazing tape as a sealant. Install the outer stops; see the BUILDING SECTION (page 158). Caulk around the windows and the bottom outside stops to prevent leaks. Add 2 × 6 blocking (and vents) or screen to enclose the rafter bays above the walls.

17 Add vertical trim at the wall corners. Trim and flash around the door opening and windows. Install flashing—and trim, if desired—along the joint where the lower roof plane meets the clerestory wall.

Aluminum flashing

18 Add 15# building paper and install the asphalt shingle roofing. The shingles should overhang the fascia drip edge by ½" along the bottom of the roof and by ⅜" along the sides. Install 1 × 4 horizontal trim boards flush with the bottom of the siding on all four walls.

19 Cut out the bottom plate inside the door's rough opening. Cut the two door panels at 30⅛ × 81". Install 1 × 4 trim around the panels, as shown in the DOOR DETAIL (page 160), using exterior wood glue and 1¼" screws or nails. Add 1 × 2 stops at the sides and top of the rough opening; see the JAMB DETAIL (page 160). Also add a 1 × 4 stop to the back side of one of the doors. Hang the doors with galvanized hinges, leaving a ¼" gap all around.

20 Finish the interior to your desired level. If you will be occupying the shed for activities, adding some wall covering, such as paneling, makes the interior much more pleasant. If you add wiring and wall insulation, the clerestory studio can function as a three-season studio in practically any climate.

Sunlight Garden Shed

This unique outbuilding is part greenhouse and part shed, making it perfect for a year-round garden space or backyard sunroom, or even an artist's studio. The front facade is dominated by windows—four 29 × 72-inch windows on the roof, plus four 29 × 18-inch windows on the front wall. When appointed as a greenhouse, two long planting tables inside the shed let you water and tend to plants without flooding the floor. If gardening isn't in your plans, you can omit the tables and cover the entire floor with plywood, or perhaps fill in between the floor timbers with pavers or stones.

Some other details that make this 10 × 12-foot shed stand out are the homemade Dutch door, with top and bottom halves that you can open together or independently, and its traditional saltbox shape. The roof covering shown here consists of standard asphalt shingles, but cedar shingles make for a nice upgrade.

Because sunlight plays a central role in this shed design, consider the location and orientation carefully. To avoid shadows from nearby structures, maintain a distance between the shed and any structure that's at least 2½ times the height of the obstruction. With all of that sunlight, the temperature inside the shed is another important consideration. You may want to install some roof vents to release hot air and water vapor.

Building the sunlight garden shed involves a few unconventional construction steps. First, the side walls are framed in two parts: You build the square portion of the end walls first, then move onto the roof framing. After the rafters are up, you complete the "rake," or angled, sections of the side walls. This makes it easy to measure for each wall stud, rather than having to calculate the lengths beforehand. Second, the shed's 4 × 4 floor structure also serves as its foundation. The plywood floor decking goes on after the walls are installed, rather than before.

Enjoy your hobbies or your plants in the attractive sunlight garden shed.

SHOPPING LIST

DESCRIPTION	QTY/SIZE	MATERIAL
FOUNDATION/FLOOR		
Foundation base & interior drainage beds	5 cu. yds.	Compactible gravel
Floor joists & blocking	7 @ 10'	4 × 4 pressure-treated landscape timbers
4 × 4 blocking	1 @ 10', 1 @ 8'	4 × 4 pressure-treated landscape timbers
Box sills (rim joists)	2 @ 12'	2 × 4 pressure treated
Nailing cleats & 2 × 4 blocking	2 @ 8'	2 × 4 pressure treated
Floor sheathing	2 sheets @ 4 × 8'	¾" ext.-grade plywood
WALL FRAMING		
Bottom plates	2 @ 12', 2 @ 10'	2 × 4 pressure treated
Top plates	4 @ 12', 2 @ 10'	2 × 4
Studs	43 @ 8'	2 × 4
Door header & jack studs	3 @ 8'	2 × 4
Rafter header	2 @ 12'	2 × 8
ROOF FRAMING		
Rafters—A & C, & nailers	10 @ 12'	2 × 4
Rafters—B & lookouts	10 @ 10'	2 × 4
Ridge board	1 @ 14'	2 × 6
EXTERIOR FINISHES		
Rear fascia	1 @ 14'	1 × 6 cedar
Rear soffit	1 @ 14'	1 × 8 cedar
Gable fascia (rake board) & soffit	4 @ 16'	1 × 6 cedar
Siding	10 sheets @ 4 × 8'	⅝" Texture 1-11 plywood siding
Siding flashing	10 linear ft.	Metal Z-flashing
Trim*	4 @ 12', 1 @ 12'	1 × 4 cedar, 1 × 2 cedar
Wall corner trim	6 @ 8'	1 × 4 cedar
ROOFING		
Sheathing	5 sheets @ 4 × 8'	½" exterior-grade plywood roof sheathing
15# building paper	1 roll	
Drip edge	72 linear ft.	Metal drip edge
Shingles	2⅔ squares	Asphalt shingles— 250# per sq. min.

DESCRIPTION	QTY/SIZE	MATERIAL
WINDOWS		
Glazing	4 pieces @ 31¼ × 76½", 4 pieces @ 31¼ × 20¾"	¼"-thick clear plastic glazing
Window stops	12 @ 10'	2 × 4
Glazing tape	60 linear ft.	
Clear exterior caulk	5 tubes	
DOOR		
Trim & stops	3 @ 8'	1 × 2 cedar
Surround	4 @ 8'	2 × 2 cedar
Z-flashing	3 linear ft.	
PLANT TABLES (OPTIONAL)		
Front table, top & trim	6 @ 12'	1 × 6 cedar or pressure treated
Front table, plates & legs	4 @ 12'	2 × 4 pressure treated
Rear table, top & trim	6 @ 8'	1 × 6 cedar or pressure treated
Rear table, plates & legs	4 @ 8'	2 × 4 pressure treated
FASTENERS & HARDWARE		
16d galvanized common nails	5 lbs.	
16d common nails	16 lbs.	
10d common nails	1½ lbs.	
8d galvanized common nails	2 lbs.	
8d galvanized box nails	3 lbs.	
10d galvanized finish nails	2½ lbs.	
8d galvanized siding nails	8 lbs.	
1" galvanized roofing nails	7 lbs.	
8d galvanized casing nails	3 lbs.	
6d galvanized casing nails	2 lbs.	
Door hinges with screws	4 @ 3½"	Corrosion-resistant hinges
Door handle	1	
Sliding bolt latch	1	
Construction adhesive	1 tube	
2" galvanized screws	1 lb.	
Door hinges with screws	6 @ 3½"	
Door handle	2	
Door lock (optional)	1	

*NOTE: The 1 × 4 trim bevel at the bottom of the sloped windows can be steeper (45° or more) so the trim slopes away from the window if there is concern that the trim may capture water running down the glazing (see WINDOW DETAIL, page 176).

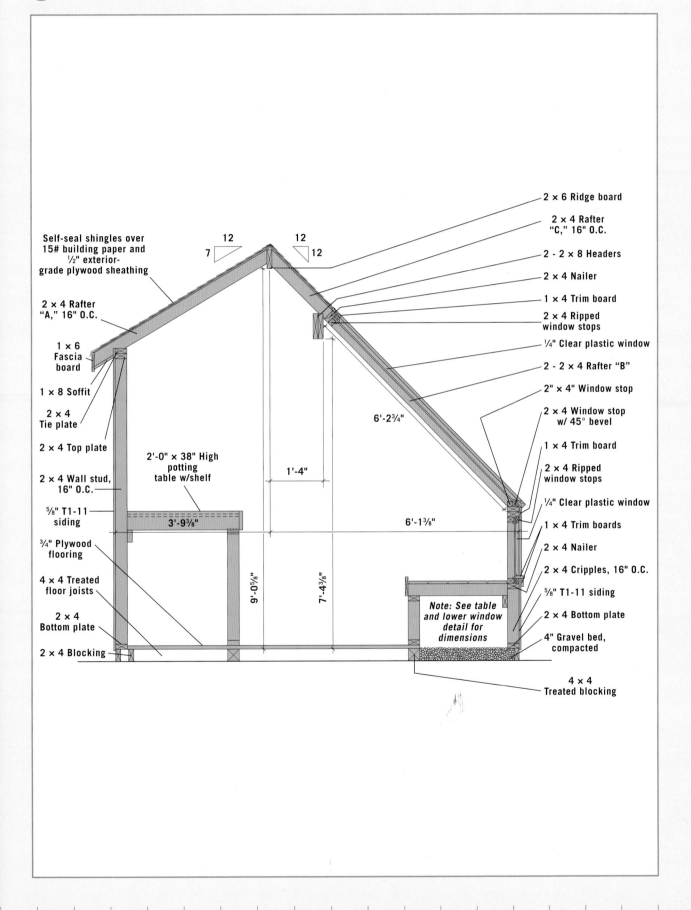

Self-seal shingles over 15# building paper and ½" exterior-grade plywood sheathing

2 × 4 Rafter "A," 16" O.C.

1 × 6 Fascia board

1 × 8 Soffit

2 × 4 Tie plate

2 × 4 Top plate

2 × 4 Wall stud, 16" O.C.

⅝" T1-11 siding

¾" Plywood flooring

4 × 4 Treated floor joists

2 × 4 Bottom plate

2 × 4 Blocking

2'-0" × 38" High potting table w/shelf

3'-9⅜"

12
7

12
12

1'-4"

6'-2¾"

6'-1⅜"

9'-0⅝"

7'-4⅜"

Note: See table and lower window detail for dimensions

2 × 6 Ridge board

2 × 4 Rafter "C," 16" O.C.

2 - 2 × 8 Headers

2 × 4 Nailer

1 × 4 Trim board

2 × 4 Ripped window stops

¼" Clear plastic window

2 - 2 × 4 Rafter "B"

2" × 4" Window stop

2 × 4 Window stop w/ 45° bevel

1 × 4 Trim board

2 × 4 Ripped window stops

¼" Clear plastic window

1 × 4 Trim boards

2 × 4 Nailer

2 × 4 Cripples, 16" O.C.

⅝" T1-11 siding

2 × 4 Bottom plate

4" Gravel bed, compacted

4 × 4 Treated blocking

FLOOR FRAMING PLAN

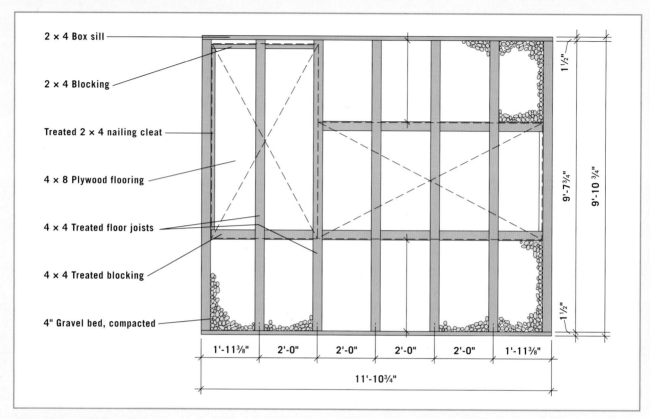

2 × 4 Box sill

2 × 4 Blocking

Treated 2 × 4 nailing cleat

4 × 8 Plywood flooring

4 × 4 Treated floor joists

4 × 4 Treated blocking

4" Gravel bed, compacted

1½"

9'-7¾"

9'-10 ¾"

1½"

1'-11⅜" 2'-0" 2'-0" 2'-0" 2'-0" 1'-11⅜"

11'-10¾"

LEFT FRAMING

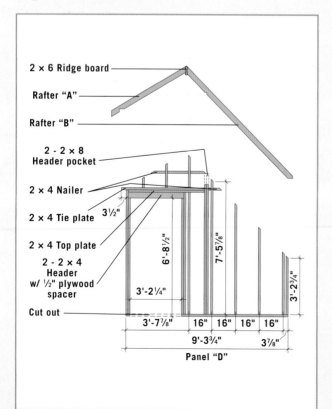

2 × 6 Ridge board

Rafter "A"

Rafter "B"

2 - 2 × 8 Header pocket

2 × 4 Nailer

2 × 4 Tie plate

2 × 4 Top plate

2 - 2 × 4 Header w/ ½" plywood spacer

Cut out

3½"

6'-8½"

7'-5⅞"

3'-2¼"

3'-2¾"

3'-7⅞" 16" 16" 16" 16"

9'-3¾" 3⅞"

Panel "D"

RIGHT FRAMING

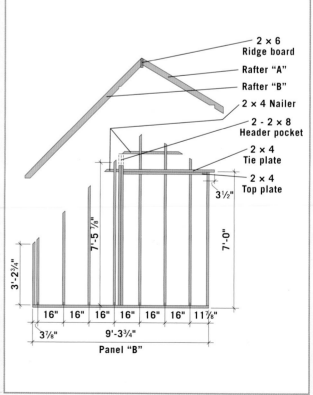

2 × 6 Ridge board

Rafter "A"

Rafter "B"

2 × 4 Nailer

2 - 2 × 8 Header pocket

2 × 4 Tie plate

2 × 4 Top plate

3½"

7'-5⅞"

7'-0"

3'-2¾"

3⅞" 16" 16" 16" 16" 16" 16" 11⅞"

9'-3¾"

Panel "B"

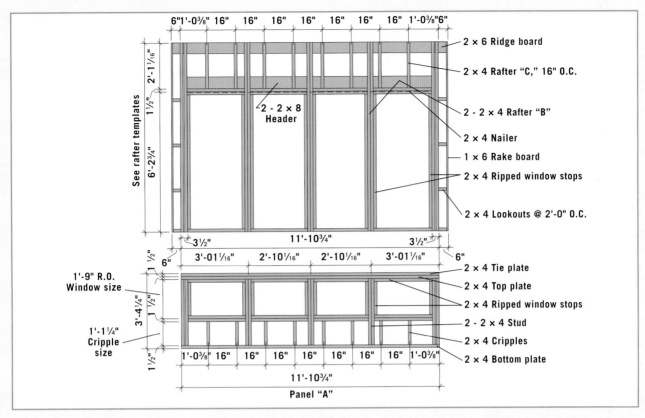

6" 1'-0⅜" 16" 16" 16" 16" 16" 16" 16" 1'-0⅜"6"

2 × 6 Ridge board

2 × 4 Rafter "C," 16" O.C.

2 - 2 × 4 Rafter "B"

2 × 4 Nailer

1 × 6 Rake board

2 × 4 Ripped window stops

2 × 4 Lookouts @ 2'-0" O.C.

See rafter templates

2'-1¹⁄₁₆"
1½"
6'-2¾"

2 - 2 × 8 Header

3½"
11'-10¾"
3½"

½"
6"
3'-01¹⁄₁₆"
2'-10¹⁄₁₆"
2'-10¹⁄₁₆"
3'-01¹⁄₁₆"
6"

1'-9" R.O. Window size

2 × 4 Tie plate

2 × 4 Top plate

2 × 4 Ripped window stops

2 - 2 × 4 Stud

2 × 4 Cripples

2 × 4 Bottom plate

1
½"
1½"
3'-4¼"
1
½"
1½"

1'-1¼" Cripple size

1'-0⅜" 16" 16" 16" 16" 16" 16" 1'-0⅜"

11'-10¾"

Panel "A"

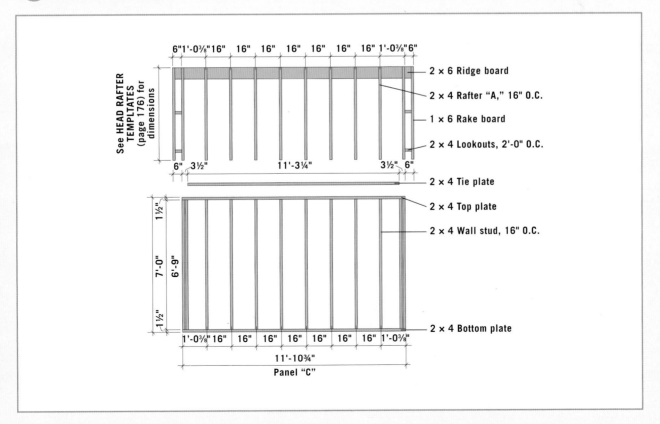

6" 1'-0⅜" 16" 16" 16" 16" 16" 16" 1'-0⅜"6"

2 × 6 Ridge board

2 × 4 Rafter "A," 16" O.C.

1 × 6 Rake board

2 × 4 Lookouts, 2'-0" O.C.

2 × 4 Tie plate

2 × 4 Top plate

2 × 4 Wall stud, 16" O.C.

2 × 4 Bottom plate

See HEAD RAFTER TEMPLATES (page 176) for dimensions

6" 3½" 11'-3¼" 3½" 6"

1½"
7'-0"
6'-9"
1½"

1'-0⅜" 16" 16" 16" 16" 16" 16" 1'-0⅜"

11'-10¾"

Panel "C"

FRONT ELEVATION

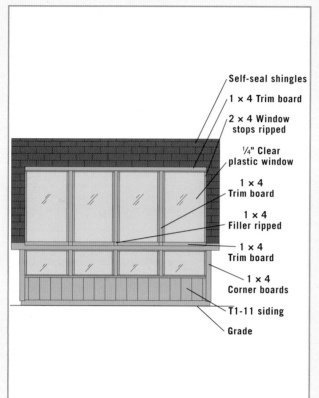

- Self-seal shingles
- 1 × 4 Trim board
- 2 × 4 Window stops ripped
- ¼" Clear plastic window
- 1 × 4 Trim board
- 1 × 4 Filler ripped
- 1 × 4 Trim board
- 1 × 4 Corner boards
- T1-11 siding
- Grade

REAR ELEVATION

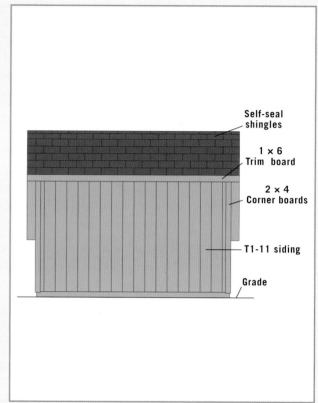

- Self-seal shingles
- 1 × 6 Trim board
- 2 × 4 Corner boards
- T1-11 siding
- Grade

RIGHT ELEVATION

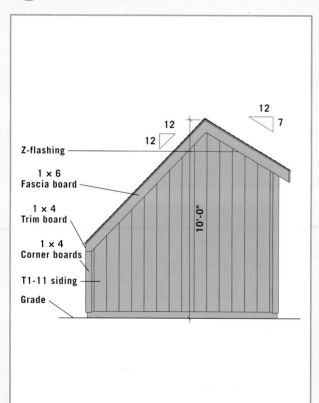

- 12 / 12
- 12 / 12
- 12 / 7
- Z-flashing
- 1 × 6 Fascia board
- 1 × 4 Trim board
- 1 × 4 Corner boards
- T1-11 siding
- Grade
- 10'-0"

SOFFIT DETAIL

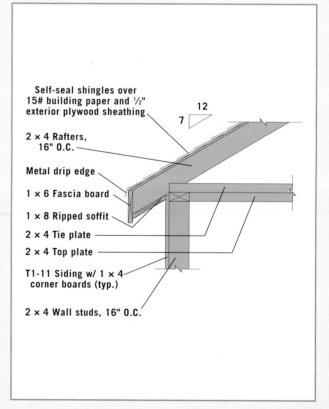

- Self-seal shingles over 15# building paper and ½" exterior plywood sheathing
- 2 × 4 Rafters, 16" O.C.
- Metal drip edge
- 1 × 6 Fascia board
- 1 × 8 Ripped soffit
- 2 × 4 Tie plate
- 2 × 4 Top plate
- T1-11 Siding w/ 1 × 4 corner boards (typ.)
- 2 × 4 Wall studs, 16" O.C.
- 7 / 12

FRONT & SIDE DOOR CONSTRUCTION

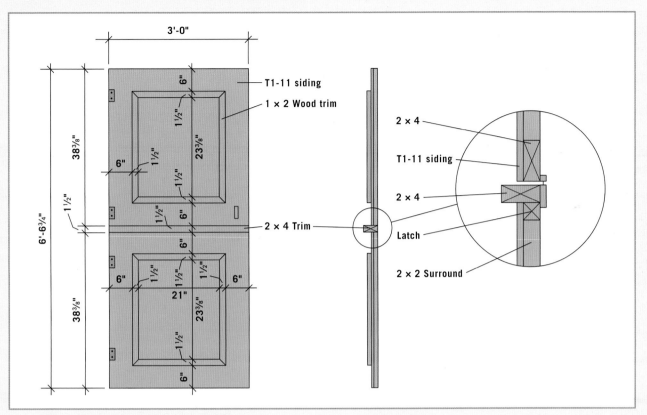

3'-0"

T1-11 siding

1 × 2 Wood trim

6"

1½"

23⅜"

1½"

38⅜"

6"

1½"

1½"

1½"

1½"

6"

2 × 4 Trim

6'-6¾"

1½"

1½"

1½"

6"

6"

38⅜"

21"

23⅜"

1½"

6"

2 × 4

T1-11 siding

2 × 4

Latch

2 × 2 Surround

FRONT & SIDE DOOR CONSTRUCTION (DOOR JAMB, REAR, DOOR HEADER)

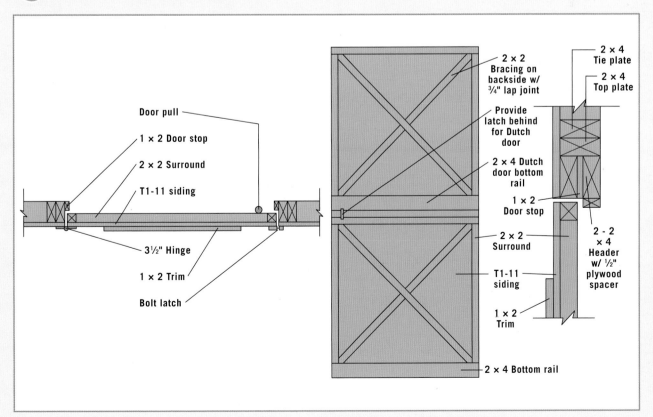

Door pull

1 × 2 Door stop

2 × 2 Surround

T1-11 siding

3½" Hinge

1 × 2 Trim

Bolt latch

2 × 2 Bracing on backside w/ ¾" lap joint

Provide latch behind for Dutch door

2 × 4 Dutch door bottom rail

1 × 2 Door stop

2 × 2 Surround

T1-11 siding

1 × 2 Trim

2 × 4 Bottom rail

2 × 4 Tie plate

2 × 4 Top plate

2 - 2 × 4 Header w/ ½" plywood spacer

HEADER & WINDOW DETAIL

2 - 2 × 4 Rafters

Self-seal shingles over 15# building paper and ½" exterior plywood sheathing

Z-flashing

1 × 4 Trim board

2 × 4 Nailer

2 - 2 × 8 Header glued and nailed

2 × 4 Ripped window stop

¼" Clear plastic window panel

2 × 4 Ripped window stop with caulking

12

12

WINDOW SECTION

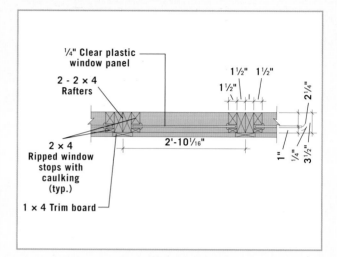

¼" Clear plastic window panel

2 - 2 × 4 Rafters

2 × 4 Ripped window stops with caulking (typ.)

1 × 4 Trim board

1½" 1½"

1½"

2¼"

2'-10¹/₁₆"

1"

¼"

3½"

WINDOW DETAIL

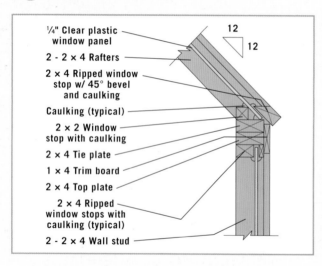

¼" Clear plastic window panel

2 - 2 × 4 Rafters

2 × 4 Ripped window stop w/ 45° bevel and caulking

Caulking (typical)

2 × 2 Window stop with caulking

2 × 4 Tie plate

1 × 4 Trim board

2 × 4 Top plate

2 × 4 Ripped window stops with caulking (typical)

2 - 2 × 4 Wall stud

12

12

TABLE & LOWER WINDOW DETAIL

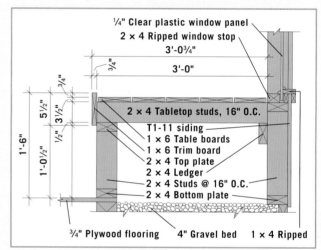

¼" Clear plastic window panel

2 × 4 Ripped window stop

3'-0¾"

3'-0"

¾"

¾"

3½"

½"

5½"

1'-0½"

1'-6"

2 × 4 Tabletop studs, 16" O.C.

T1-11 siding

1 × 6 Table boards

1 × 6 Trim board

2 × 4 Top plate

2 × 4 Ledger

2 × 4 Studs @ 16" O.C.

2 × 4 Bottom plate

¾" Plywood flooring   4" Gravel bed   1 × 4 Ripped

HEAD RAFTER TEMPLATES

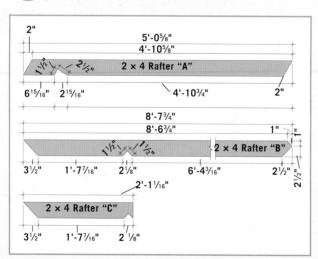

2"

5'-0⁵/₈"

4'-10⁵/₈"

1½"

2½"

2 × 4 Rafter "A"

6¹⁵/₁₆"   2¹⁵/₁₆"   4'-10¾"   2"

8'-7¾"

8'-6¾"   1"

1½"   1½"

2 × 4 Rafter "B"

3½"   1'-7⁷/₁₆"   2⅛"   6'-4³/₁₆"   2½"   2½"

2'-1¹/₁₆"

2 × 4 Rafter "C"

3½"   1'-7⁷/₁₆"   2⅛"

RAKE BOARD DETAIL

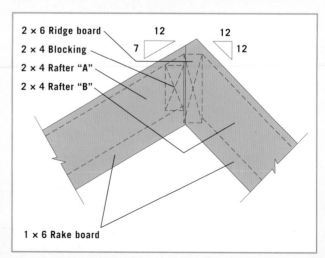

2 × 6 Ridge board

2 × 4 Blocking

2 × 4 Rafter "A"

2 × 4 Rafter "B"

12

7

12

12

1 × 6 Rake board

How to Build the Sunlight Garden Shed

Cut ten **4 × 4 blocks** to fit between the joists. Install six blocks 34½" from the front rim joist, and install four blocks 31½" from the rear. Toenail the blocks to the joists. All blocks, joists, and sills must be flush at the top.

Build the foundation, following the basic steps used for a wooden skid foundation (page 14). First, prepare a bed of compacted gravel. Make sure the bed is flat and level. Cut seven 4 × 4" × 10-ft. pressure-treated posts down to 115¾" to serve as floor joists. Position the joists as shown in the FLOOR FRAMING PLAN (page 172). Level each joist, and make sure all are level with one another and the ends are flush. Add rim joists and blocking: Cut two 12-ft. 2 × 4s (142¾") for rim joists. Fasten the rim joists to the ends of the 4 × 4 joists (see the FLOOR FRAMING PLAN) with 16d galvanized common nails.

To frame the rear wall, cut one top plate and one pressure-treated bottom plate (142¾"). Cut twelve studs (81"). Assemble the wall following the layout in REAR FRAMING (page 173). Raise the wall and fasten it to the rear rim joist and the intermediate joists, using 16d galvanized common nails. Brace the wall in position with 2 × 4 braces staked to the ground.

For the front wall, cut two top plates and one treated bottom plate (142¾"). Cut ten studs (35¾") and eight cripple studs (13¼"). Cut four 2 × 4 window sills (31¹¹⁄₁₆"). Assemble the wall following the layout in the FRONT FRAMING (page 173). Add the double top plate, but do not install the window stops at this time. Raise, attach, and brace the front wall. *(continued)*

SHED PROJECTS ● 177

Cut lumber for the right side wall: one top plate (54⅞"), one treated bottom plate (111¾"), four studs (81"), and two header post studs (86⅞"); and for the left side wall: top plate (54⅞"), bottom plate (111¾"), three studs (81"), two jack studs (77½"), two posts (86⅞"), and a built-up 2 × 4 header (39¼"). Assemble and install the walls as shown in RIGHT SIDE FRAMING and LEFT SIDE FRAMING (page 172). Add the doubled top plates along the rear and side walls. Install treated 2 × 4 nailing cleats to the joists and blocking as shown in the FLOOR FRAMING PLAN (page 172) and BUILDING SECTION (page 171).

Trim two sheets of ¾" plywood as needed and install them over the joists and blocking as shown in the FLOOR FRAMING PLAN (page 172), leaving open cavities along the front of the shed and a portion of the rear. Fasten the sheets with 8d galvanized common nails driven every 6" along the edges and 8" in the field. Fill the exposed foundation cavities with 4" of crushed gravel and compact it thoroughly.

Construct the rafter header from two 2 × 8s cut to 142¾". Join the pieces with construction adhesive and pairs of 10d common nails driven every 24" on both sides. Set the header on top of the sidewall posts, and toenail it to the posts with four 16d common nails at each end.

Cut one of each "A" and "B" pattern rafters using the HEAD RAFTER TEMPLATES (page 176). Test-fit the rafters. The B rafter should rest squarely on the rafter header, and its bottom end should sit flush with outside of the front wall. Adjust the rafter cuts as needed, then use the pattern rafters to mark and cut the remaining A and B rafters.

Cut the 2 × 6 ridge board (154¾"). Mark the rafter layout onto the ridge and front and rear wall plates following FRONT FRAMING and REAR FRAMING (page 173). Install the A and B rafters and ridge. Make sure the B rafters are spaced accurately so the windows will fit properly into their frames; see the WINDOW SECTION (page 176).

Cut a pattern "C" rafter, test-fit, and adjust as needed. Cut the remaining seven C rafters and install them. Measure and cut four 2 × 4 nailers (31¹¹⁄₁₆") to fit between the sets of B rafters (as shown). Position the nailers as shown in the HEADER & WINDOW DETAIL (page 176) and toenail them to the rafters.

Complete the rake portions of each side wall. Mark the stud layouts onto the bottom plate, and onto the top plate of the square wall section; see RIGHT and LEFT SIDE FRAMING (page 172). Use a plumb bob to transfer the layout to the rafters. Measure for each stud, cutting the top ends of the studs under the B rafters at 45° and those under the A rafters at 30°. Toenail the studs to the plates and rafters. Add horizontal 2 × 4 nailers as shown in the framing drawings.

Blade guard removed for clarity

Create the inner and outer window stops from 10-ft.-long 2 × 4s. For stops at the sides and tops of the roof windows and all sides of the front wall windows, rip the inner stops to 2¼" wide and the outer stops to 1" wide; see the WINDOW SECTION and WINDOW DETAIL (page 176). For the bottom of each roof window, rip the inner stop to 1½"; bevel the edge of the outer stop at 45°.

(continued)

Install each window as follows: Attach inner stops as shown in the drawings, using galvanized finish nails. Paint or varnish the rafters and stops for moisture protection. Apply a heavy bead of caulk at each location shown on the drawings (HEADER & WINDOW DETAIL, WINDOW SECTION/DETAIL, TABLE & LOWER WINDOW DETAIL, page 176). Set the glazing in place, add another bead of caulk, and attach the outer stops. Cover the rafters and stop edges with 1 × 4 trim.

Cover the walls with T1-11 siding, starting with the rear wall. Trim the sheets as needed so they extend from the bottom edges of the rafters down to at least 1" below the tops of the foundation timbers. On the side walls, add Z-flashing above the first row and continue the siding up to the rafters.

Install 1 × 6 fascia over the ends of the A rafters. Keep all fascia ½" above the rafters so it will be flush with the roof sheathing. Using scrap rafter material, cut the 2 × 4 lookouts (5¼"). On each outer B rafter, install one lookout at the bottom end and four more spaced 24" on-center going up. On the A rafters, add a lookout at both ends and two spaced evenly in between. Install the 1 × 6 rake boards (fascia) as shown in the RAKE BOARD DETAIL (page 176).

Soffit

Rip 1 × 6 boards to 5¼" width (some may come milled to 5¼" already) for the gable soffits. Fasten the soffits to the lookouts with siding nails. Rip a 1 × 8 board for the soffit along the rear eave, beveling the edges at 30° to match the A rafter ends. Install the soffit.

Deck the roof with ½" plywood sheathing, starting at the bottom ends of the rafters. Install the/a metal drip edge, building paper, and asphalt shingles. If desired, add one or more roof vents during the shingle installation. Be sure to overlap shingles onto the 1 × 4 trim board above the roof windows, as shown in the HEADER & WINDOW DETAIL (page 176).

Construct the planting tables from 2 × 4 lumber and 1 × 6 boards, as shown in the TABLE & LOWER WINDOW DETAIL (page 176) and BUILDING SECTION (page 171). The bottom plates of the table legs should be flush with the outside edges of the foundation blocking.

Build each of the two door panels using T1-11 siding, 2 × 2 bracing, a 2 × 4 bottom rail, and 1 × 2 trim on the front; see the DOOR CONSTRUCTION drawings (page 175). The panels are identical except for a 2 × 4 sill added to the top of the lower panel. Install 1 × 2 stops at the sides and top of the door opening. Hang the doors with four hinges, leaving even gaps all around. Install a bolt latch for locking the two panels together.

Complete the trim details with 1 × 4 vertical corner boards, 1 × 4 horizontal trim above the front wall windows, and ripped 1 × 4 trim and 1 × 2 trim at the bottom of the front wall windows (see the TABLE & LOWER WINDOW DETAIL, page 176). Paint the siding and trim, or coat with exterior wood finish.

Gambrel Garage

Following classic barn designs, this 12 × 12-foot garage-sized storage shed has several features that make it a versatile storage shed or workshop. The garage's 144-square-foot floor is a poured concrete slab with a thickened edge that allows it to serve as the building's foundation. Designed for economy and durability, the floor can easily support heavy machinery, woodworking tools, and recreational vehicles.

The garage's sectional overhead door makes for quick access to equipment and supplies and provides plenty of air and natural light for working inside. The door opening is sized for an 8-foot-wide × 7-foot-tall door, but you can buy any size or style of door you like—just make your door selection before you start framing the garage.

Another important design feature of this building is its gambrel roof, which maximizes the usable interior space. Beneath the roof is a sizeable storage attic with 315 cubic feet of space and its own double doors above the garage door.

THE GAMBREL ROOF

The gambrel roof is the defining feature of two structures in American architecture: the barn and the Dutch Colonial house. Adopted from earlier English buildings, the gambrel style became popular in America during the early 17th century and was used on homes and farm buildings throughout the Atlantic region. Today, the gambrel roof remains a favorite detail for designers of sheds, garages, and carriage houses.

The basic gambrel shape has two flat planes on each side, with the lower plane sloped much more steeply than the upper. More elaborate versions incorporate a flared eave, known as a "Dutch kick," that was often extended to shelter the front and rear facades of the building. Barns typically feature an extended peak at the front, sheltering the doors of the hayloft. The main advantage of the gambrel roof is the increased space underneath the roof, providing additional headroom for upper floors in homes or extra storage space in outbuildings.

All that space comes at a price—construction complexity. The roof rafters (twice as many as most other roofs) are cut at different angles. Mix up your cuts and you can find yourself with a lot of waste lumber. Even if you get all the angles correct, you'll be doing a lot of work with your saw.

But there is an easier way. Manufacturers have developed a solution to multiple rafter cuts for a gambrel roof. These innovative and relatively inexpensive options allow you to use lumber cut to length and nailed into the prefab brackets that create the angles automatically. Count the number of rafters you'll be building for the gambrel shed design you've chosen, and then just order the appropriate number of brackets. Then simply assemble your rafter trusses and install as described in this project. They're virtually foolproof and will save you a great deal of time, and possibly frustration and lumber costs as well!

SHOPPING LIST

DESCRIPTION	QTY/SIZE	MATERIAL
FOUNDATION		
Drainage material	1.75 cu. yds.	Compactible gravel
Concrete slab	2.5 cu. yds.	3,000 psi concrete
Mesh	144 sq. ft.	6 × 6", W1.4 × W1.4 welded wire mesh
WALL FRAMING		
Bottom plates	4 @ 12'	2 × 4 pressure treated
Top plates	8 @ 12'	2 × 4
Studs	47 @ 92⅝"	2 × 4
Headers	2 @ 10', 2 @ 6'	2 × 8
Header spacers	1 @ 9', 1 @ 6'	½" plywood—7" wide
Angle braces	1 @ 4'	2 × 4
GABLE WALL FRAMING		
Plates	2 @ 10'	2 × 4
Studs	7 @ 10'	2 × 4
Header	2 @ 6'	2 × 6
Header spacer	1 @ 5'	½" plywood—5" wide
ATTIC FLOOR		
Joists	10 @ 12'	2 × 6
Floor sheathing	3 sheets @ 4 × 8'	¾" tongue-&-groove ext.-grade plywood
KNEEWALL FRAMING		
Bottom plates	2 @ 12'	2 × 4
Top plates	4 @ 12'	2 × 4
Studs	8 @ 10'	2 × 4
Nailers	2 @ 14'	2 × 8
ROOF FRAMING		
Rafters	28 @ 10'	2 × 4
Metal anchors—rafters	20, with nails	Simpson H2.5
Collar ties	2 @ 6'	2 × 4
Ridge board	1 @ 14'	2 × 6
Lookouts	1 @ 10'	2 × 4
Soffit ledgers	2 @ 14'	2 × 4
Soffit blocking	6 @ 8'	2 × 4
EXTERIOR FINISHES		
Plywood siding	14 sheets @ 4 × 8'	⅝" Texture 1-11 plywood, grooves 8" ≈O. C.
Z-flashing—siding	2 pieces @ 12'	Galvanized 18-gauge
Horizontal wall trim	2 @ 12'	1 × 4 cedar
Corner trim	8 @ 8'	1 × 4 cedar
Fascia	6 @ 10', 2 @ 8'	1 × 6 cedar
Subfascia	4 @ 8'	1 × 4 pine
Plywood soffits	1 sheet @ 10'	⅜" cedar or fir plywood
Soffit vents	4 @ 4 × 12"	Louver w/bug screen
Z-flashing—garage door	1 @ 10'	Galvanized 18-gauge

DESCRIPTION	QTY/SIZE	MATERIAL
ROOFING		
Roof sheathing	12 sheets @ 4 × 8'	½" plywood
Shingles	3 squares	250# per square (min.)
15# building paper	300 sq. ft.	
Metal drip edge	2 @ 14', 2 @ 12'	Galvanized metal
Roof vents (optional)	2 units	
WINDOW		
Frame	3 @ 6'	¾ × 4" (actual) S4S cedar
Stops	4 @ 8'	1 × 2 S4S cedar
Glazing tape	30 linear ft.	
Glass	1 piece—field measure	¼" clear, tempered
Exterior trim	3 @ 6'	1 × 4 S4S cedar
Interior trim (optional)	3 @ 6'	1 × 2 S4S cedar
DOOR		
Frame	3 @ 8'	1 × 6 S4S cedar
Door sill	1 @ 6'	1 × 6 S4S cedar
Stops	1 @ 8', 1 @ 6'	1 × 2 S4S cedar
Panel material	4 @ 8'	1 × 8 T&G V-joint S4S cedar
Door X-brace/panel trim	4 @ 6', 2 @ 8'	1 × 4 S4S cedar
Exterior trim	1 @ 8', 1 @ 6'	1 × 4 S4S cedar
Interior trim (optional)	1 @ 8', 1 @ 6'	1 × 2 S4S cedar
Strap hinges	4	
GARAGE DOOR		
Frame	3 @ 8'	1 × 8 S4S cedar
Door	1 @ 8' × 6' - 8"	Sectional flush door w/2" track
Rails	2 @ 8'	2 × 6
Trim	3 @ 8'	1 × 4 S4S cedar
FASTENERS		
Anchor bolts	16	⅜" × 8", with washers & nuts, galvanized
16d galvanized common nails	2 lbs.	
16d common nails	17 lbs.	
10d common nails	2 lbs.	
10d galvanized casing nails	1 lb.	
8d common nails	3 lbs.	
8d galvanized finish nails	6 lbs.	
8d box nails	6 lbs.	
6d galvanized finish nails	20 nails	
3d galvanized box nails	½ lb.	
⅞" galvanized roofing nails	2½ lbs.	
2½" deck screws	24 screws	
1¼" wood screws	48 screws	
Construction adhesive	2 tubes	
Silicone-latex caulk	2 tubes	

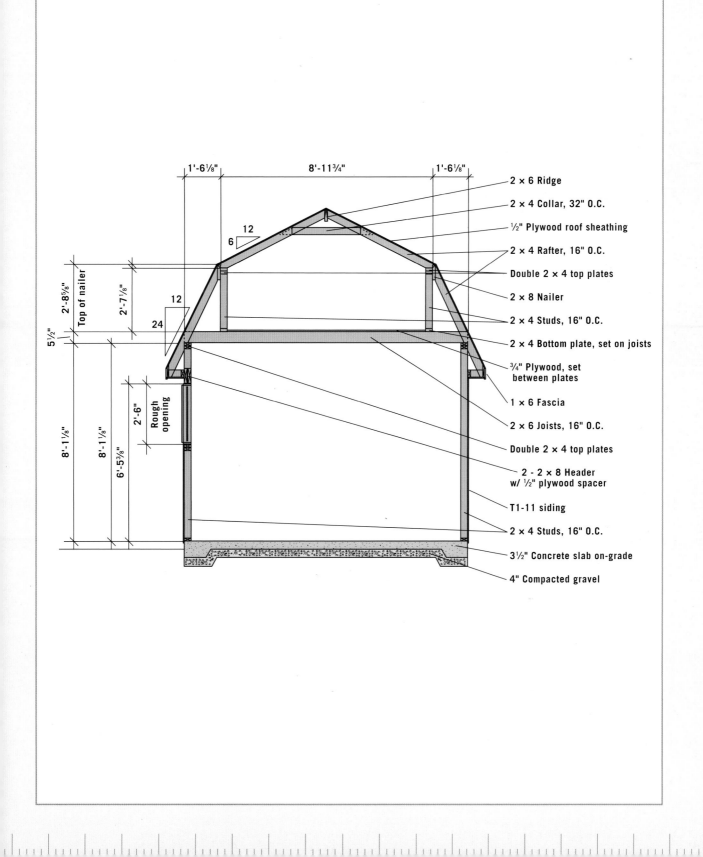

Labels (top to bottom, right side):
- 2 × 6 Ridge
- 2 × 4 Collar, 32" O.C.
- ½" Plywood roof sheathing
- 2 × 4 Rafter, 16" O.C.
- Double 2 × 4 top plates
- 2 × 8 Nailer
- 2 × 4 Studs, 16" O.C.
- 2 × 4 Bottom plate, set on joists
- ¾" Plywood, set between plates
- 1 × 6 Fascia
- 2 × 6 Joists, 16" O.C.
- Double 2 × 4 top plates
- 2 - 2 × 8 Header w/ ½" plywood spacer
- T1-11 siding
- 2 × 4 Studs, 16" O.C.
- 3½" Concrete slab on-grade
- 4" Compacted gravel

Dimensions:
- 1'-6⅛"
- 8'-11¾"
- 1'-6⅛"
- 12 / 6
- 12 / 24
- 2'-8⅝"
- Top of nailer
- 2'-7⅛"
- 5½"
- 8'-1⅛"
- 8'-1⅛"
- 6'-5⅜"
- 2'-6"
- Rough opening

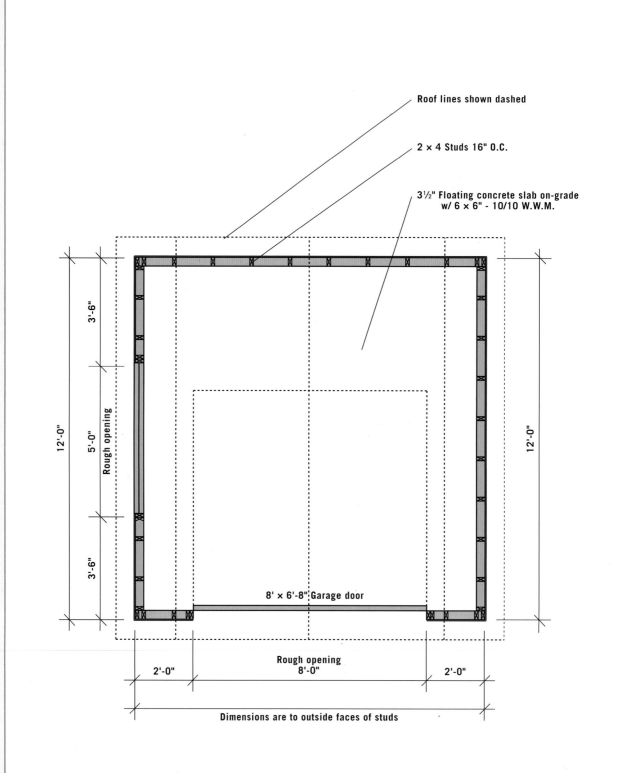

Roof lines shown dashed

2 × 4 Studs 16" O.C.

3½" Floating concrete slab on-grade
w/ 6 × 6" - 10/10 W.W.M.

12'-0"

3'-6"

5'-0"
Rough opening

3'-6"

12'-0"

8' × 6'-8" Garage door

Rough opening
8'-0"

2'-0"

2'-0"

Dimensions are to outside faces of studs

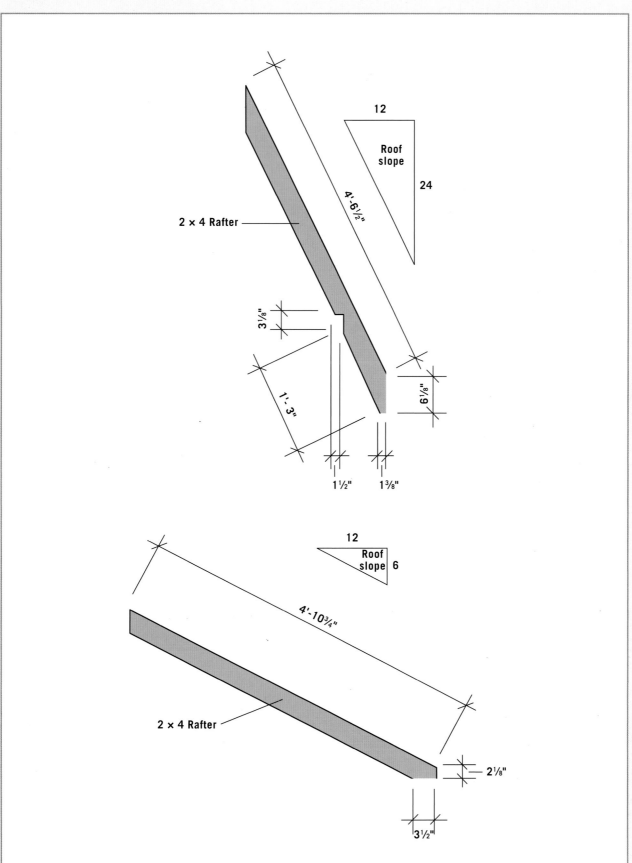

12
Roof slope
24

2 × 4 Rafter

4'-6½"

3⅛"

1'-3"

6⅛"

1½"   1⅜"

12
Roof slope 6

4'-10¾"

2 × 4 Rafter

2⅛"

3½"

FRONT ELEVATION

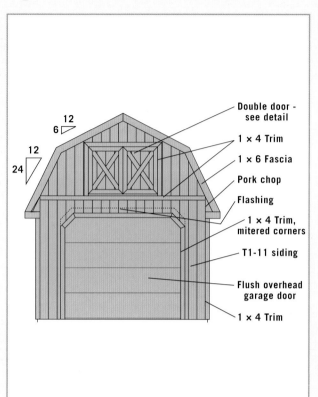

12
6

12
24

Double door - see detail

1 × 4 Trim

1 × 6 Fascia

Pork chop

Flashing

1 × 4 Trim, mitered corners

T1-11 siding

Flush overhead garage door

1 × 4 Trim

LEFT ELEVATION

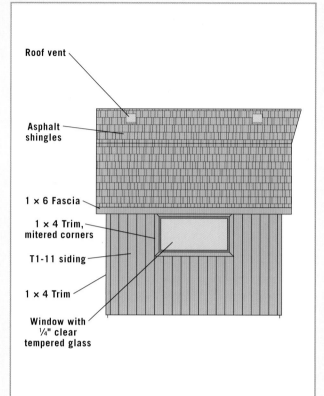

Roof vent

Asphalt shingles

1 × 6 Fascia

1 × 4 Trim, mitered corners

T1-11 siding

1 × 4 Trim

Window with ¼" clear tempered glass

REAR ELEVATION

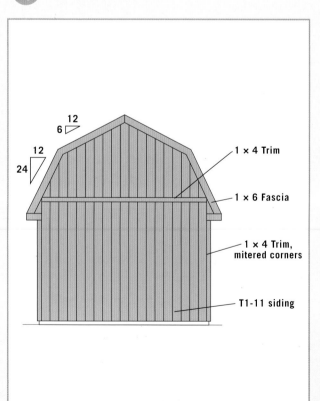

12
6

12
24

1 × 4 Trim

1 × 6 Fascia

1 × 4 Trim, mitered corners

T1-11 siding

RIGHT ELEVATION

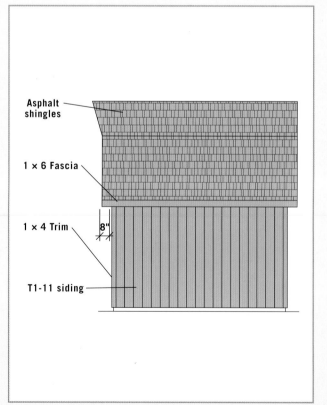

Asphalt shingles

1 × 6 Fascia

1 × 4 Trim

8"

T1-11 siding

GABLE OVERHANG DETAIL

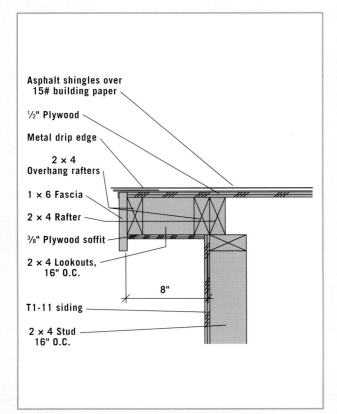

Asphalt shingles over
15# building paper

½" Plywood

Metal drip edge

2 × 4
Overhang rafters

1 × 6 Fascia

2 × 4 Rafter

⅜" Plywood soffit

2 × 4 Lookouts,
16" O.C.

8"

T1-11 siding

2 × 4 Stud
16" O.C.

GABLE OVERHANG RAFTER DETAILS

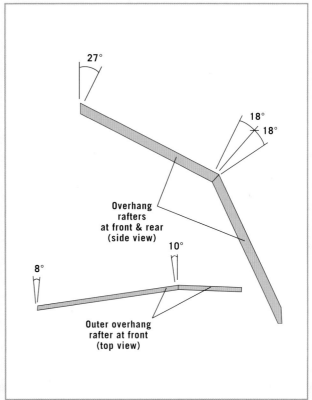

27°

18°
18°

Overhang
rafters
at front & rear
(side view)

10°

8°

Outer overhang
rafter at front
(top view)

EAVE DETAIL

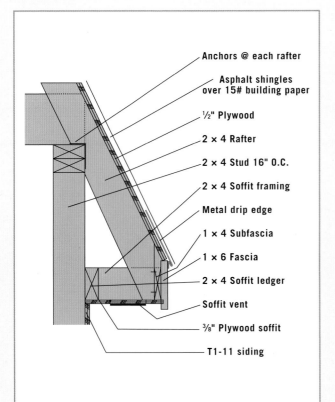

Anchors @ each rafter

Asphalt shingles
over 15# building paper

½" Plywood

2 × 4 Rafter

2 × 4 Stud 16" O.C.

2 × 4 Soffit framing

Metal drip edge

1 × 4 Subfascia

1 × 6 Fascia

2 × 4 Soffit ledger

Soffit vent

⅜" Plywood soffit

T1-11 siding

SILL DETAIL

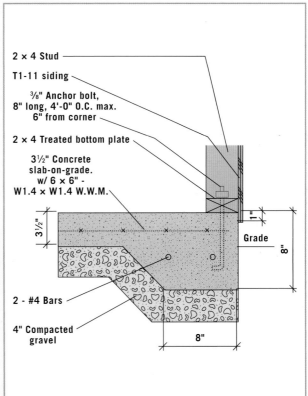

2 × 4 Stud

T1-11 siding

⅜" Anchor bolt,
8" long, 4'-0" O.C. max.
6" from corner

2 × 4 Treated bottom plate

3½" Concrete
slab-on-grade.
w/ 6 × 6" -
W1.4 × W1.4 W.W.M.

3½"

1"

Grade

8"

2 - #4 Bars

4" Compacted
gravel

8"

ATTIC DOOR ELEVATION

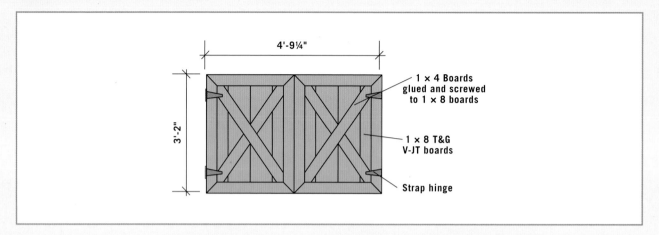

4'-9¼"

3'-2"

1 × 4 Boards
glued and screwed
to 1 × 8 boards

1 × 8 T&G
V-JT boards

Strap hinge

ATTIC DOOR JAMB DETAIL

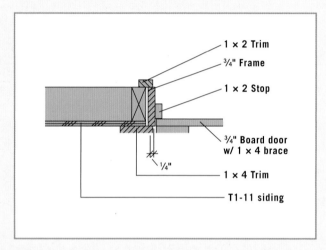

1 × 2 Trim

¾" Frame

1 × 2 Stop

¾" Board door
w/ 1 × 4 brace

¼"

1 × 4 Trim

T1-11 siding

GARAGE DOOR TRIM DETAIL

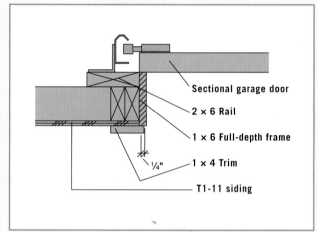

Sectional garage door

2 × 6 Rail

1 × 6 Full-depth frame

¼"

1 × 4 Trim

T1-11 siding

ATTIC DOOR SILL DETAIL

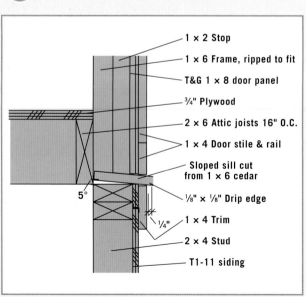

1 × 2 Stop

1 × 6 Frame, ripped to fit

T&G 1 × 8 door panel

¾" Plywood

2 × 6 Attic joists 16" O.C.

1 × 4 Door stile & rail

Sloped sill cut
from 1 × 6 cedar

5°

⅛" × ⅛" Drip edge

¼"

1 × 4 Trim

2 × 4 Stud

T1-11 siding

WINDOW JAMB DETAIL

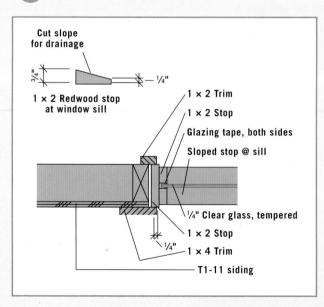

Cut slope
for drainage

¾"

¼"

1 × 2 Redwood stop
at window sill

1 × 2 Trim

1 × 2 Stop

Glazing tape, both sides

Sloped stop @ sill

¼" Clear glass, tempered

1 × 2 Stop

¼"

1 × 4 Trim

T1-11 siding

FRONT FRAMING ELEVATION

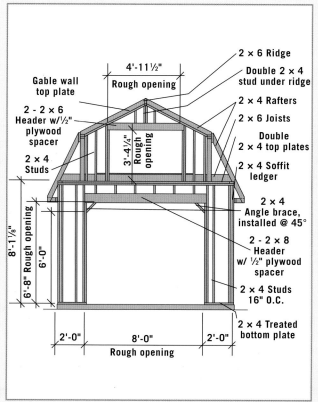

4'-11½"
Rough opening

Gable wall top plate

2 - 2 × 6 Header w/½" plywood spacer

2 × 4 Studs

3'-4¼" Rough opening

8'-1⅛" Rough opening

6'-8" Rough opening

6'-0"

2'-0"

8'-0"
Rough opening

2'-0"

2 × 6 Ridge

Double 2 × 4 stud under ridge

2 × 4 Rafters

2 × 6 Joists

Double 2 × 4 top plates

2 × 4 Soffit ledger

2 × 4 Angle brace, installed @ 45°

2 - 2 × 8 Header w/ ½" plywood spacer

2 × 4 Studs 16" O.C.

2 × 4 Treated bottom plate

LEFT FRAMING ELEVATION

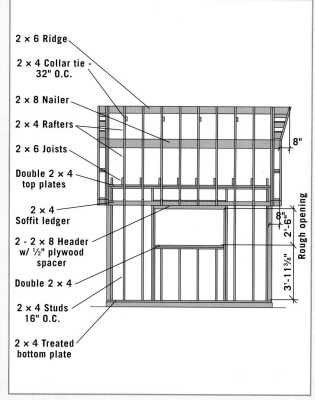

2 × 6 Ridge

2 × 4 Collar tie - 32" O.C.

2 × 8 Nailer

2 × 4 Rafters

2 × 6 Joists

Double 2 × 4 top plates

2 × 4 Soffit ledger

2 - 2 × 8 Header w/ ½" plywood spacer

Double 2 × 4

2 × 4 Studs 16" O.C.

2 × 4 Treated bottom plate

8"

8"

2'-6" Rough opening

3'-11⅜"

REAR FRAMING ELEVATION

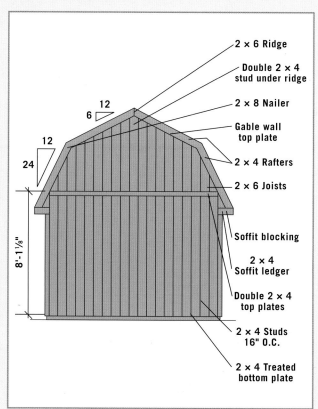

12
6

12
24

8'-1⅛"

2 × 6 Ridge

Double 2 × 4 stud under ridge

2 × 8 Nailer

Gable wall top plate

2 × 4 Rafters

2 × 6 Joists

Soffit blocking

2 × 4 Soffit ledger

Double 2 × 4 top plates

2 × 4 Studs 16" O.C.

2 × 4 Treated bottom plate

RIGHT FRAMING ELEVATION

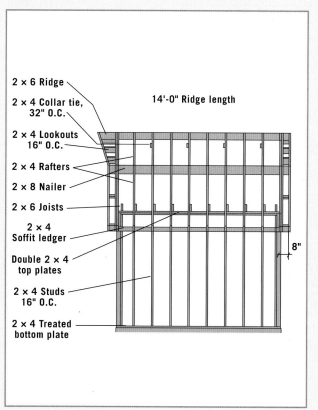

2 × 6 Ridge

2 × 4 Collar tie, 32" O.C.

2 × 4 Lookouts 16" O.C.

2 × 4 Rafters

2 × 8 Nailer

2 × 6 Joists

2 × 4 Soffit ledger

Double 2 × 4 top plates

2 × 4 Studs 16" O.C.

2 × 4 Treated bottom plate

14'-0" Ridge length

8"

How to Build the Gambrel Garage

(1)

Build the slab foundation at 144 × 144", following the steps on page 21. Set J-bolts into the concrete 1¾" from the outer edges and extending 2½" from the surface. Set a bolt 6" from each corner and every 48" in between (except in the door opening). Let the slab cure for at least three days before you begin construction.

Snap chalk lines on the slab for the wall plates. Cut two bottom plates and two top plates at 137" for the sidewalls. Cut two bottom and two top plates at 144" for the front and rear walls. Use pressure-treated lumber for all bottom plates. Cut 38 studs at 92⅝", plus two jack studs for the garage door at 78½" and two window studs at 75⅞".

NOTE: Add the optional slab now, if desired.

(2)

(3)

Construct the built-up 2 × 8 headers at 99" (garage door) and 63" (window). Frame, install, and brace the walls with double top plates one at a time, following the FLOOR PLAN (page 186) and ELEVATION drawings (page 188). Use galvanized nails to attach the studs to the sole plates. Anchor the walls to the J-bolts in the slab with galvanized washers and nuts.

(4)

Build the attic floor. Cut ten 2 × 6 joists to 144" long, then clip each top corner with a 1½"-long, 45° cut. Install the joists as shown in FRAMING ELEVATIONS (page 191), leaving a 3½" space at the front and rear walls for the gable wall studs. Fasten the joists with three 8d nails at each end.

Frame the attic kneewalls: Cut four top plates at 144" and two bottom plates at 137". Cut 20 studs at 26⅝" and four end studs at 33⅝". Lay out the plates so the studs fall over the attic joists. Frame the walls and install them 18⅛" from the ends of the joists, then add temporary bracing.

OPTION: You can begin building the roof frame by cutting two 2 × 8 nailers to 144" long. Fasten the nailers to the kneewalls so their top edges are 32⅝" above the attic joists.

Cover the attic floor between the kneewalls with ¾" plywood. Run the sheets perpendicular to the joists, and stop them flush with the outer joists. Fasten the flooring with 8d ring-shank nails every 6" along the edges and every 12" in the field of the sheets.

Mark the rafter layouts onto the top and outside faces of the 2 × 8 nailers; see the FRAMING ELEVATIONS (page 191).

Cut the 2 × 6 ridge board at 168", mitering the front end at 16°. Mark the rafter layout onto the ridge. The outer common rafters should be 16" from the front end and 8" from the rear end of the ridge. *(continued)*

Use the **RAFTER TEMPLATES** (page 187) to mark and cut two upper pattern rafters and one lower pattern rafter. Test-fit the rafters and make any needed adjustments. Use the patterns to mark and cut the remaining common rafters (20 total of each type). For the gable overhangs, cut an additional eight lower and six upper rafters following the GABLE OVERHANG RAFTER DETAILS (page 189).

Install the common rafters; then reinforce the joints at the knee walls with framing connectors. Also nail the attic joists to the sides of the floor rafters. Cut four 2 × 4 collar ties at 34", mitering the ends at 26.5°. Fasten them between pairs of upper rafters, as shown in the BUILDING SECTION (page 185) and FRAMING ELEVATIONS (page 191).

Snap a chalk line across the sidewall studs, level with the ends of the rafters. Cut two 2 × 4 soffit ledgers at 160" and fasten them to the studs on top of the chalk lines, with their ends overhanging the walls by 8". Cut 24 2 × 4 blocks to fit between the ledger and rafter ends, as shown in the EAVE DETAIL (page 189). Install the blocks.

Frame the gable overhangs. Cut 12 2 × 4 lookouts at 5" and nail them to the inner overhang rafters as shown in the LEFT and RIGHT SIDE FRAMING ELEVATIONS (page 188). Install the inner overhang rafters over the common rafters, using 10d nails. Cut the two front (angled) overhang rafters; see the GABLE OVERHANG RAFTER DETAILS (page 189). Install those rafters; then add two custom-cut lookouts for each rafter.

To complete the gable walls, cut top plates to fit between the ridge and the attic kneewalls. Install the plates flush with the outer common rafters. Mark the stud layout onto the walls and gable top plate; see the FRONT and REAR FRAMING ELEVATIONS (page 188). Cut the gable studs to fit and install them. Construct the built-up 2 × 6 attic door header at 62½"; then clip the top corners to match the roof slope. Install the header with jack studs cut at 40¼".

Install siding on the walls, holding it 1" below the top of the concrete slab. Add Z-flashing along the top edges, and then continue the siding up to the rafters. Below the attic door opening, stop the siding about ¼" below the top wall plate, as shown in the ATTIC DOOR SILL DETAIL (page 190). Don't nail the siding to the garage door header until the flashing is installed (Step 20).

Mill a ⅜"-wide × ¼"-deep groove into the 1 × 6 boards for the horizontal fascia along the eaves and gable ends (about 36 linear ft.); see the EAVE DETAIL (page 189). Use a router or table saw with a dado-head blade to mill the groove, and make the groove ⅞" above the bottom edge of the fascia.

Install the 1 × 4 subfascia along the eaves, keeping the bottom edge flush with the ends of the rafters and the ends flush with the outsides of the outer-most rafters; see the EAVE DETAIL (page 189). Add the milled fascia at the eaves, aligning the top of the groove with the bottom of the subfascia. Cut fascia to wrap around the overhangs at the gable ends but don't install them until Step 17.   *(continued)*

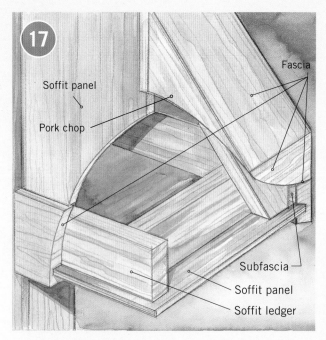

Add fascia at the gable ends, holding it up ½" to be flush with the roof sheathing. Cut soffit panels to fit between the fascia and walls, and fasten them with 3d galvanized nails. Install the end and return fascia pieces at the gable overhangs. Enclose each overhang at the corners with a triangular piece of grooved fascia (called a pork chop) and a piece of soffit material. Install the soffit vents as shown in the EAVE DETAIL (page 189).

Sheath the roof, starting at one of the lower corners. Add metal drip edge along the eaves, followed by building paper; then add drip edge along the gable ends, over the paper. Install the asphalt shingles. Plan the courses so the roof transition occurs midshingle, not between courses; the overlapping shingles will relax over time. If desired, add roof vents.

Cover the Z-flashing at the rear wall with horizontal 1 × 4 trim. Finish the four wall corners with overlapping vertical 1 × 4 trim. Install the 2 × 6 rails that will support the garage door tracks, following the door manufacturer's instructions to determine the sizing and placement; see the GARAGE DOOR TRIM DETAIL (page 190).

For the garage doorframe, rip 1 × 8 trim boards to width so they cover the front wall siding and 2 × 6 rails, as shown in the GARAGE DOOR TRIM DETAIL (page 190). Install the trim, mitering the pieces at 22.5°. Install the 1 × 4 trim around the outside of the opening, adding flashing along the top; see the FRONT ELEVATION (page 188).

Install the garage door in the door opening, following the manufacturer's directions.

Build the window frame, which should be ½" narrower and shorter than the rough opening. Install the frame using shims and 10d galvanized casing nails, as shown in the WINDOW JAMB DETAIL (page 190). Cut eight 1 × 2 stop pieces to fit the frame. Bevel the outer sill stop for drainage. Order glass to fit, or cut your own plastic panel. Install the glazing and stops, using glazing tape for a watertight seal. Add the window trim.

For the attic doorframe, rip 1 × 6s to match the depth of the opening and cut the head jamb and side jambs. Cut the sill from full-width 1 × 6 stock; then cut a kerf for a drip edge (see the ATTIC DOOR SILL DETAIL, page 190). Fasten the head jamb to the side jambs and install the sill at a 5° slope between the side jambs. Install the doorframe using shims and 10d casing nails. Add shims or cedar shingles along the length of the sill to provide support underneath. The front edge of the frame should be flush with the face of the siding. Add 1 × 2 stops at the frame sides and top, ¾" from the front edges.

Build the attic doors as shown in the ATTIC DOOR ELEVATION (page 190), using glue and 1¼" screws. Each door measures 28⅝ × 38", including the panel braces. Cut the 1 × 8 panel boards about ⅛" short along the bottom to compensate for the sloping sill. Install the door panels with two hinges each. Add 1 × 4 horizontal trim on the front wall, up against the doorsill; then trim around both sides of the doorframe. Prime and paint as desired.

Convenience Shed

The convenience shed is so named for its exceptional versatility and ample storage space. This classic gabled outbuilding has a footprint that measures 12 × 16 feet and it includes several features not found in most storage sheds. For starters, its 8-foot-wide overhead garage door provides easy access for large equipment, supplies, projects or even a small automobile. The foundation and shed floor is a poured concrete slab, so it's ideal for heavy items such as lawn tractors and stationary tools.

To the right of the garage door is a box bay window. This special architectural detail gives the building's facade a surprising house-like quality while filling the interior with natural light. And the bay's 33-inch-deep × 60-inch-wide sill platform is the perfect place for herb pots or an indoor flower box. The adjacent wall includes a second large window and a standard service door, making this end of the shed a pleasant, convenient space for all kinds of work or leisure.

Above the main space of the convenience shed is a fully framed attic built with 2 × 6 joists for supporting plenty of stored goods. The steep pitch of the roof allows for over 3 feet of headroom under the peak. Access to the attic is provided by a drop-down staircase that folds up and out of the way, leaving the workspace clear below.

The garage door, service door, staircase, and both windows of the shed are pre-built factory units that you install following the manufacturers' instructions. Be sure to order all of the units before starting construction. This makes it easy to adjust the framed openings, if necessary, to match the precise sizing of each unit. Also consult your local building department to learn about design requirements for the concrete foundation. You may need to extend and/or reinforce the perimeter portion of the slab or include a footing that extends below the frost line. An extended apron (as seen in Gambrel Garage, page 182) is very useful if you intend to house vehicles in the shed.

Get creative in this large shed that can pass for a small garage. The building's ample space and generous headroom offer endless options, including a practice site for your shed band that's well out of earshot from the house.

SHOPPING LIST

DESCRIPTION	QTY/SIZE	MATERIAL
FOUNDATION		
Drainage material	2.75 cu. yd.	Compactible gravel
Concrete slab	Field measure	3,000 psi concrete
Mesh	200 sq. ft.	6 × 6", W1.4 × W1.4 welded wire mesh
Reinforcing bar	As required by local code	As required by local code
WALL FRAMING		
Bottom plates	1 @ 16', 2 @ 12', 1 @ 10'	2 × 4 pressure treated
Top plates	2 @ 14', 4 @ 12', 4 @ 10'	2 × 4
Standard wall studs	51 @ 8'* *may use 92⅝" precut studs	2 × 4
Diagonal bracing	5 @ 12'	1 × 4 (std. lumber)
Jack studs	5 @ 14'	2 × 4
Gable end studs	5 @ 8'	2 × 4
Header, overhead door	2 @ 10'	2 × 12
Header, windows	2 @ 10'	2 × 12
Header, service door	1 @ 8'	2 × 12
Header & stud spacers		See Sheathing, below
BOX BAY FRAMING		
Half-wall bottom plate	1 @ 8'	2 × 4 pressure treated
Half-wall top plate & studs	3 @ 8'	2 × 4
Joists	3 @ 8'	2 × 6
Window frame	4 @ 12'	2 × 4
Sill platform & top	1 sheet @ 4 × 8'	½" plywood
Rafter blocking	1 @ 8'	2 × 8
ROOF FRAMING		
Rafters (& lookouts, blocking)	36 @ 10'	2 × 6
Ridge board	1 @ 18'	2 × 8
ATTIC		
Floor joists	16 @ 12'	2 × 6
Floor decking	6 sheets @ 4 × 8'	½" plywood
Staircase	1 unit for 22 × 48" rough opening	Disappearing attic stair unit

DESCRIPTION	QTY/SIZE	MATERIAL
EXTERIOR FINISHES		
Eave fascia	2 @ 18'	2 × 8 cedar
Gable fascia	4 @ 10'	1 × 8 cedar
Drip edge & gable trim	160 linear ft.	1 × 2 cedar
Siding	15 sheets @ 4 × 8'	⅝" Texture 1-11 plywood siding w/ vertical grooves 8" on center (or similar)
Siding flashing	30 linear ft.	Metal Z-flashing
Overhead door jambs	1 @ 10', 2 @ 8'	1 × 6 cedar
Overhead door stops	3 @ 8'	Cedar door stop
Overhead door surround	1 @ 10', 2 @ 8'	2 × 6
Corner trim	8 @ 8'	1 × 4 cedar
Door & window trim	4 @ 8', 5 @ 10'	1 × 4 cedar
Box bay bottom trim	1 @ 8'	1 × 10 cedar
ROOFING		
Sheathing (& header, stud spacers)	14 sheets @ 4 × 8'	½" exterior-grade plywood roof sheathing
15# building paper	2 rolls	
Shingles	4⅔ squares	Asphalt shingles—250# per sq. min.
Roof flashing	10'6"	
DOORS & WINDOWS		
Overhead garage door w/ hardware	1 @ 8' × 7'	
Service door	1 unit for 38 × 72⅞" rough opening	Prehung exterior door unit
Window	2 units for 57 × 41⅜"	Casement mullion window unit—complete
FASTENERS & HARDWARE		
J-bolts w/ nuts & washers	14	½"-dia. × 12"
16d galvanized common nails	3 lbs.	
16d common nails	15 lbs.	
10d common nails	2½ lbs.	
8d box nails	16 lbs.	
8d common nails	5 lbs.	
8d galvanized siding nails	10 lbs.	
1" galvanized roofing nails	10 lbs.	
8d galvanized casing nails	3 lbs.	
Entry door lockset	1	

FOUNDATION PLAN

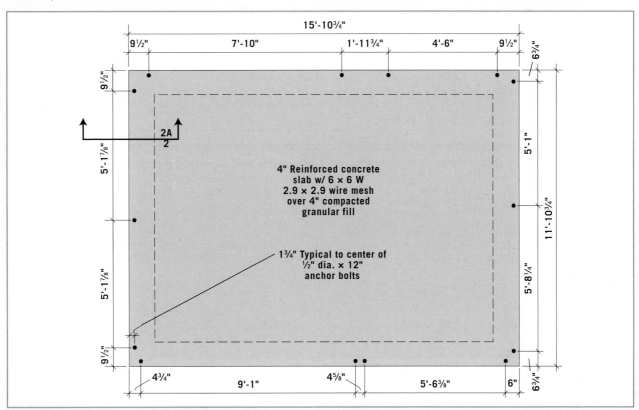

15'-10¾"

9½" 7'-10" 1'-11¾" 4'-6" 9½" 6¾"

9½"

5'-1⅞"

2A / 2

4" Reinforced concrete
slab w/ 6 × 6 W
2.9 × 2.9 wire mesh
over 4" compacted
granular fill

5'-1"

11'-10¾"

5'-1⅞"

1¾" Typical to center of
½" dia. × 12"
anchor bolts

5'-8¼"

9½"

4¾" 9'-1" 4⅝" 5'-6⅜" 6" 6¾"

FOUNDATION DETAIL

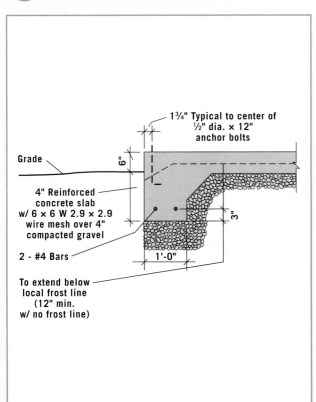

1¾" Typical to center of
½" dia. × 12"
anchor bolts

6"

Grade

4" Reinforced
concrete slab
w/ 6 × 6 W 2.9 × 2.9
wire mesh over 4"
compacted gravel

2 - #4 Bars

3"

1'-0"

To extend below
local frost line
(12" min.
w/ no frost line)

BUILDING SECTION

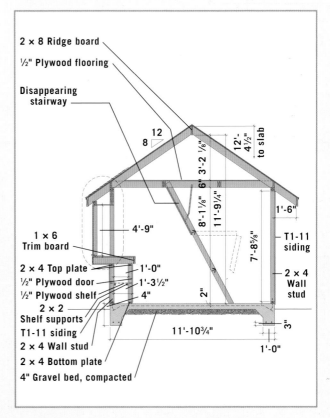

2 × 8 Ridge board

½" Plywood flooring

Disappearing
stairway

12
8

12'-
4½"
to slab

3'-2⅛"

6"

6"

8'-11⅛"

11'-9¼"

1'-6"

1 × 6
Trim board

4'-9"

T1-11
siding

2 × 4 Top plate
½" Plywood door
½" Plywood shelf
2 × 2
Shelf supports
T1-11 siding
2 × 4 Wall stud
2 × 4 Bottom plate
4" Gravel bed, compacted

1'-0"
1'-3½"
4"

2"

7'-8⅝"

2 × 4
Wall
stud

11'-10¾"

3"

1'-0"

FRONT ELEVATION

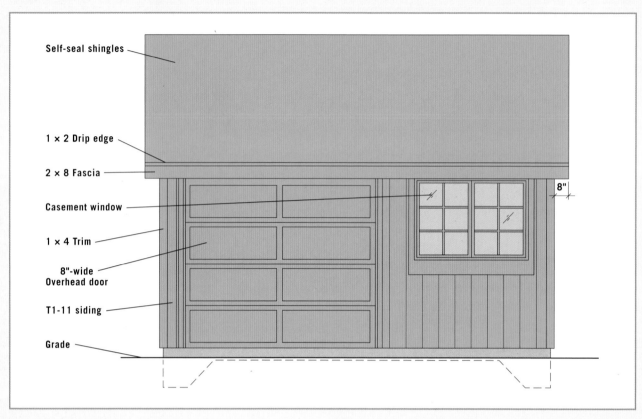

Self-seal shingles

1 × 2 Drip edge

2 × 8 Fascia

Casement window

1 × 4 Trim

8"-wide
Overhead door

T1-11 siding

Grade

8"

RIGHT ELEVATION

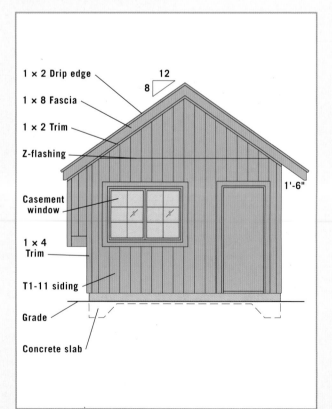

1 × 2 Drip edge

12
8

1 × 8 Fascia

1 × 2 Trim

Z-flashing

1'-6"

Casement
window

1 × 4
Trim

T1-11 siding

Grade

Concrete slab

REAR ELEVATION

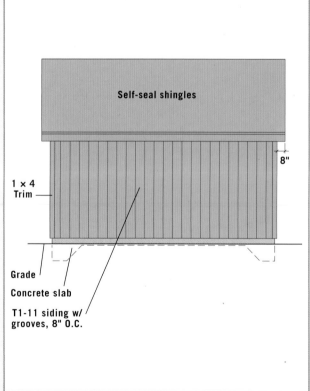

Self-seal shingles

8"

1 × 4
Trim

Grade

Concrete slab

T1-11 siding w/
grooves, 8" O.C.

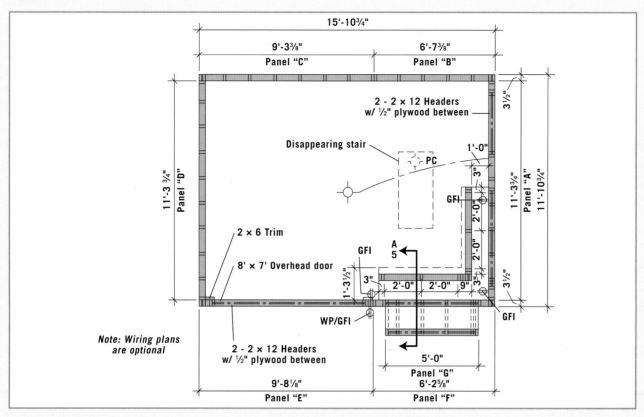

15'-10¾"

9'-3⅜"
Panel "C"

6'-7⅜"
Panel "B"

2 - 2 × 12 Headers
w/ ½" plywood between

3½"

Disappearing stair

1'-0"

PC

3"

GFI

2'-0"
2'-0"

11'-3¾"
Panel "A"
11'-10¾"

11'-3 ¾"
Panel "D"

2 × 6 Trim

GFI

A
5

2'-0"

8' × 7' Overhead door

1'-3½"

3"

2'-0" 2'-0" 9" 3"

3½"

GFI

WP/GFI

5'-0"

GFI

Note: Wiring plans
are optional

2 - 2 × 12 Headers
w/ ½" plywood between

Panel "G"
6'-2⅝"

9'-8⅛"

Panel "E"

Panel "F"

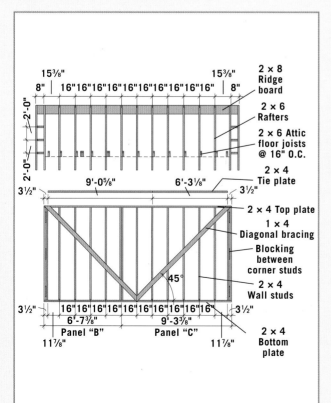

15⅜"   15⅜"
8" 16"16"16"16"16"16"16"16"16"16" 8"

2 × 8
Ridge
board

2'-0"

2 × 6
Rafters

2 × 6 Attic
floor joists
@ 16" O.C.

2'-0"

2 × 4
Tie plate

3½"

9'-0⅝"   6'-3⅛"   3½"

2 × 4 Top plate

1 × 4
Diagonal bracing

Blocking
between
corner studs

45°

2 × 4
Wall studs

3½" 16"16"16"16"16"16"16"16"16" 3½"
6'-7⅜"   9'-3⅜"

11⅞" Panel "B"   Panel "C"   11⅞"

2 × 4
Bottom
plate

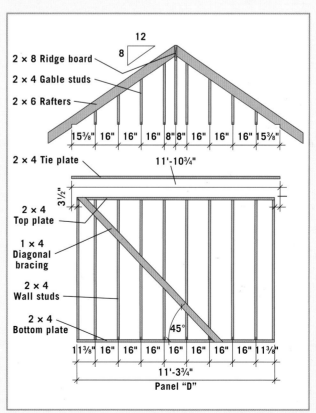

12
8

2 × 8 Ridge board

2 × 4 Gable studs

2 × 6 Rafters

15⅜" 16" 16" 16" 8" 8" 16" 16" 16" 15⅜"

2 × 4 Tie plate

11'-10¾"

3½"

2 × 4
Top plate

1 × 4
Diagonal
bracing

2 × 4
Wall studs

45°

2 × 4
Bottom plate

1⅜" 16" 16" 16" 16" 16" 16" 16" 1⅜"

11'-3¾"

Panel "D"

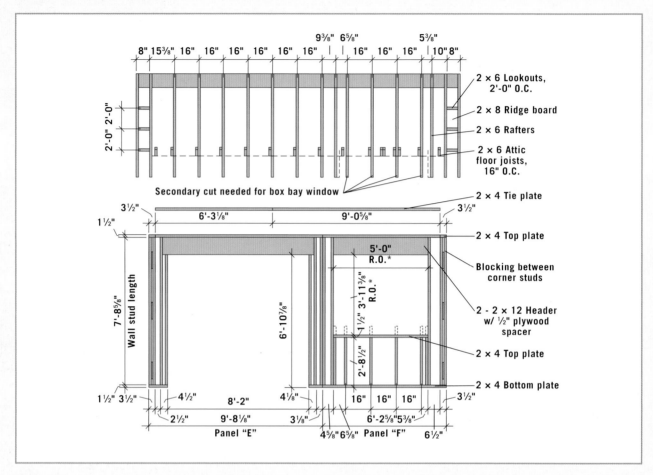

8" 15³⁄₈" 16" 16" 16" 16" 16" 16" 9³⁄₈" 6⁵⁄₈" 16" 16" 16" 5³⁄₈" 10" 8"

2 × 6 Lookouts, 2'-0" O.C.

2 × 8 Ridge board

2 × 6 Rafters

2 × 6 Attic floor joists, 16" O.C.

2'-0" 2'-0"

Secondary cut needed for box bay window

2 × 4 Tie plate

3½" 6'-3¹⁄₈" 9'-0⁵⁄₈" 3½"

1½"

2 × 4 Top plate

Blocking between corner studs

5'-0" R.O.*

3'-11³⁄₈" R.O.*

7'-8⁵⁄₈" Wall stud length

6'-10⁷⁄₈"

1½"

2'-8½"

2 - 2 × 12 Header w/ ½" plywood spacer

2 × 4 Top plate

2 × 4 Bottom plate

1½" 3½" 4½" 8'-2" 4¹⁄₈" 16" 16" 16" 3½"

2½" 9'-8¹⁄₈" 3¹⁄₈" 6'-2⁵⁄₈" 5³⁄₈"

Panel "E"   4⁵⁄₈" 6⁵⁄₈" Panel "F"   6½"

ATTIC FLOOR JOIST FRAMING

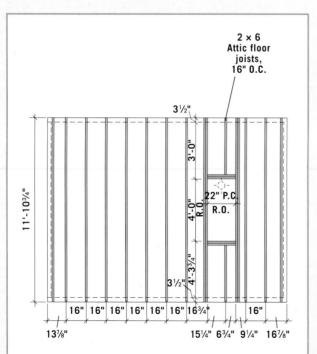

2 × 6 Attic floor joists, 16" O.C.

3½"

3'-0"

22" P.C. R.O.

4'-0"

11'-10³⁄₄"

4'-3¾"

3½"

16" 16" 16" 16" 16" 16" 16¾" 16"

13⁷⁄₈"   15¼" 6¾" 9¼" 16⁷⁄₈"

BOX BAY WINDOW FRAMING

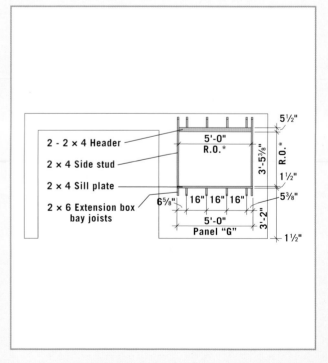

2 - 2 × 4 Header

2 × 4 Side stud

2 × 4 Sill plate

2 × 6 Extension box bay joists

5½"

5'-0" R.O.*

3'-5³⁄₈" R.O.*

1½"

6⁵⁄₈" 16" 16" 16"

5³⁄₈"

5'-0"

Panel "G"

3'-2"

1½"

OVERHEAD DOOR HEADER DETAIL

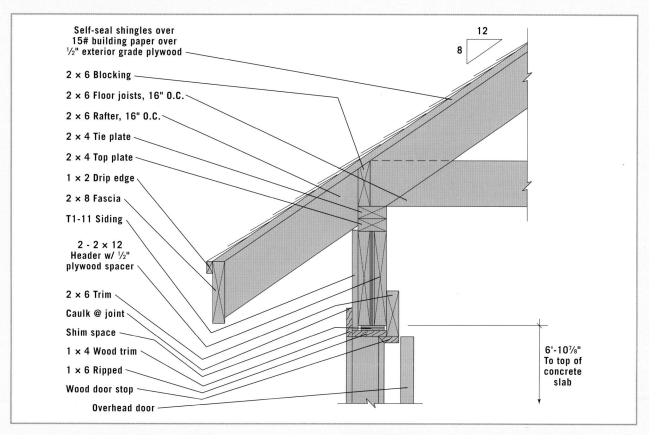

Self-seal shingles over
15# building paper over
½" exterior grade plywood

2 × 6 Blocking

2 × 6 Floor joists, 16" O.C.

2 × 6 Rafter, 16" O.C.

2 × 4 Tie plate

2 × 4 Top plate

1 × 2 Drip edge

2 × 8 Fascia

T1-11 Siding

2 - 2 × 12
Header w/ ½"
plywood spacer

2 × 6 Trim

Caulk @ joint

Shim space

1 × 4 Wood trim

1 × 6 Ripped

Wood door stop

Overhead door

12
8

6'-10⅞"
To top of
concrete
slab

OVERHEAD DOOR JAMB DETAIL

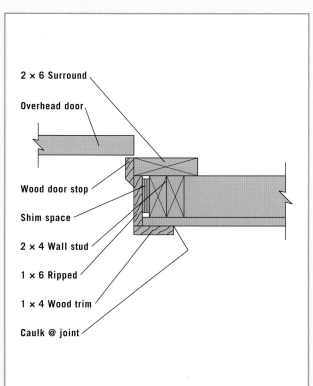

2 × 6 Surround

Overhead door

Wood door stop

Shim space

2 × 4 Wall stud

1 × 6 Ripped

1 × 4 Wood trim

Caulk @ joint

SERVICE DOOR HEADER/JAMB DETAIL

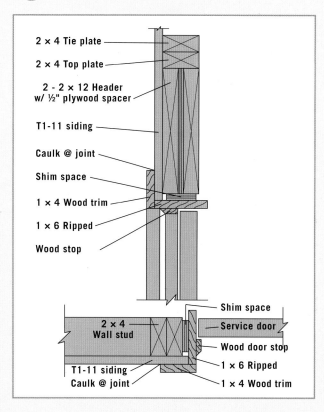

2 × 4 Tie plate

2 × 4 Top plate

2 - 2 × 12 Header
w/ ½" plywood spacer

T1-11 siding

Caulk @ joint

Shim space

1 × 4 Wood trim

1 × 6 Ripped

Wood stop

Shim space

Service door

Wood door stop

1 × 6 Ripped

1 × 4 Wood trim

2 × 4
Wall stud

T1-11 siding

Caulk @ joint

RAFTER TEMPLATE

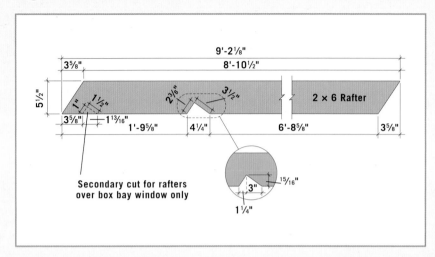

9'-2⅛"
8'-10½"
3⅝"
5½"
1"
1½"
1¹³⁄₁₆"
3⅝"
1'-9⅝"
2³⁄₈"
3½"
4¼"
6'-8⅝"
3⅝"
2 × 6 Rafter

Secondary cut for rafters
over box bay window only

15⁄16"
3"
1¼"

COUNTER DETAIL

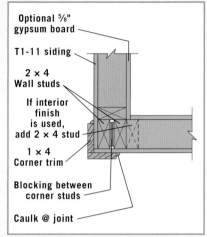

Optional ⅝"
gypsum board

T1-11 siding

2 × 4
Wall studs

If interior
finish
is used,
add 2 × 4 stud

1 × 4
Corner trim

Blocking between
corner studs

Caulk @ joint

BOX BAY WINDOW DETAIL

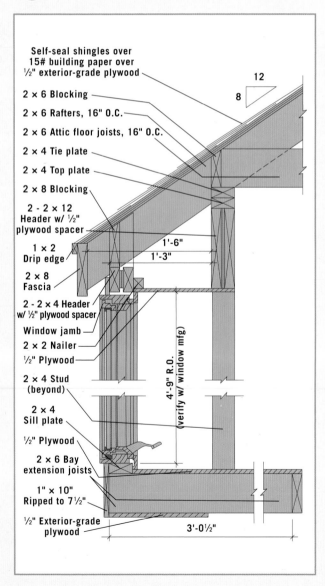

Self-seal shingles over
15# building paper over
½" exterior-grade plywood

2 × 6 Blocking

2 × 6 Rafters, 16" O.C.

2 × 6 Attic floor joists, 16" O.C.

2 × 4 Tie plate

2 × 4 Top plate

2 × 8 Blocking

2 - 2 × 12
Header w/ ½"
plywood spacer

1 × 2
Drip edge

2 × 8
Fascia

2 - 2 × 4 Header
w/ ½" plywood spacer

Window jamb

2 × 2 Nailer

½" Plywood

2 × 4 Stud
(beyond)

2 × 4
Sill plate

½" Plywood

2 × 6 Bay
extension joists

1" × 10"
Ripped to 7½"

½" Exterior-grade
plywood

12
8

1'-6"
1'-3"

4'-9" R.O.
(verify w/ window mfg)

3'-0½"

ISOMETRIC

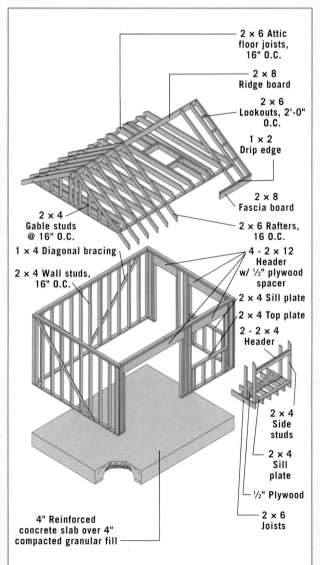

2 × 6 Attic
floor joists,
16" O.C.

2 × 8
Ridge board

2 × 6
Lookouts, 2'-0"
O.C.

1 × 2
Drip edge

2 × 8
Fascia board

2 × 6 Rafters,
16 O.C.

2 × 4
Gable studs
@ 16" O.C.

1 × 4 Diagonal bracing

2 × 4 Wall studs,
16" O.C.

4 - 2 × 12
Header
w/ ½" plywood
spacer

2 × 4 Sill plate

2 × 4 Top plate

2 - 2 × 4
Header

2 × 4
Side
studs

2 × 4
Sill
plate

½" Plywood

2 × 6
Joists

4" Reinforced
concrete slab over 4"
compacted granular fill

How to Build the Convenience Shed

Build the concrete foundation using the specifications shown in the FOUNDATION DETAIL (page 201) and following the basic procedure on pages 21 to 23. The slab should measure 190¾" × 142¾". Set the 14 J-bolts into the concrete as shown in the FOUNDATION PLAN (page 201).

NOTE: All slab specifications must comply with local building codes.

Snap chalk lines for the bottom plates so they will be flush with the outside edges of the foundation. You can frame the walls in four continuous panels or break them up into panels "A" through "F," as shown in the WALL FRAMING PLAN (page 203). We completely assembled and squared all four walls before raising and anchoring them.

Frame the rear wall(s) following the REAR FRAMING (page 203) illustration. Use pressure-treated lumber for the bottom plate, and nail it to the studs with galvanized 16d common nails. All of the standard studs are 92⅝" long. Square the wall, then add 1 × 4 let-in bracing.

Raise the rear wall and anchor it to the foundation J-bolts with washers and nuts. Brace the wall upright. Frame and raise the remaining walls one at a time, then tie all of the walls together with double top plates. Cover the outside of the walls with T1-11 siding. *(continued)*

Cut fifteen 2 × 6 attic floor joists at 142¾". Cut the top corner at both ends of each joist: Mark 1⅞" along the top edge and ¹⁵⁄₁₆" down the end; connect the marks, then cut along the line. Clipping the corner prevents the joist from extending above the rafters.

Mark the joist layout onto the wall plates following the ATTIC FLOOR JOIST FRAMING drawing (page 204). Leave 3½" between the outsides of the end walls and the outer joists. Toenail the joists to the plates with three 8d common nails at each end. Frame the rough opening for the staircase with doubled side joists and doubled headers; fasten doubled members together with pairs of 10d nails every 16". Install the drop-down staircase unit following the manufacturer's instructions.

Cover the attic floor with ½" plywood, fastening it to the joists with 8d nails.

Use the RAFTER TEMPLATE (page 206) to mark and cut two pattern rafters. Test-fit the rafters and adjust the cuts as needed. Cut all (24) standard rafters. Cut four special rafters with an extra birdsmouth cut for the box bay. Cut four gable overhang rafters—these have no birdsmouth cuts.

Cut the 2 × 8 ridge board at 206¾". Mark the rafter layout on the ridge and wall plates as shown in FRONT FRAMING (page 204) and REAR FRAMING (page 203). Frame the roof following the steps on illustrations (pages 205 to 207). Install 6½"-long lookouts 24" on center, then attach the overhang rafters. Fasten the attic joists to the rafters with three 10d nails at each end.

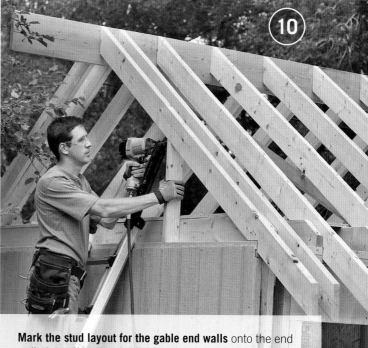

Mark the stud layout for the gable end walls onto the end wall plates following SIDE FRAMING (page 203). Transfer the layout to the rafters, using a level. Cut each of the 2 × 4 studs to fit, mitering the top ends at 33.5°. Install the studs flush with the end walls.

Construct the 2 × 4 half-wall for the interior apron beneath the box bay: Cut two plates at 60" (pressure-treated lumber for bottom plate); cut five studs at 32½". Fasten one stud at each end, and space the remaining studs evenly in between. Mark a layout line 12" from the inside of the shed's front wall (see the BUILDING SECTION page 201). Anchor the half-wall to the slab using masonry screws or a powder-actuated nailer.

Cut six 2 × 6 joists at 36½". Toenail the joists to the inner and outer half-walls following the layout in BOX BAY WINDOW FRAMING (page 204); the joists should extend 15" past the outer shed wall. Add a 60"-long 2 × 4 sill plate at the ends of the joists. Cut two 2 × 4 side studs to extend from the sill plate to the top edges of the rafters (angle top ends at 33.5°), and install them. Install a built-up 2 × 4 header between the side studs 41⅜" above the sill plate. *(continued)*

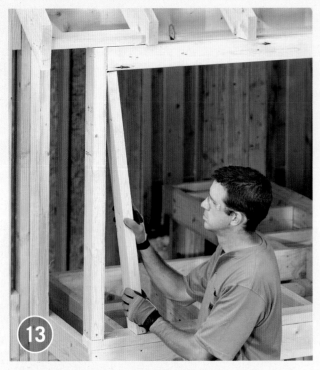

Install a 2 × 2 nailer ½" up from the bottom of the 2 × 4 bay header. Cover the top and bottom of the bay with ½" plywood as shown in the BOX BAY WINDOW DETAIL (page 206). Cut a 2 × 4 stud to fit between the plywood panels at each end of the 2 × 4 shed wall header; fasten these to the studs supporting the studs and the header.

Bevel the side edge of the 2 × 6 blocking stock at 33.5°. Cut individual blocks to fit between the rafters and attic joists, and install them to seal off the rafter bays; see the OVERHEAD DOOR HEADER DETAIL (page 205). The blocks should be flush with the tops of the rafters. Custom-cut 2 × 8 blocking to enclose the rafter bays above the box bay header; see the BOX BAY WINDOW DETAIL (page 206).

Add 2 × 8 fascia to the ends of the rafters along each eave so the top outer edge will be flush with the top of the roof sheathing. Cover the gable overhang rafters with 1 × 8 fascia. Add 1 × 2 trim to serve as a drip edge along the eaves and gable ends so it will be flush with the top of the roof sheathing.

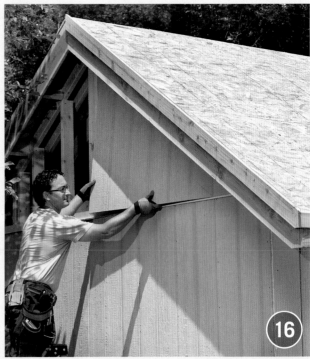

Add Z-flashing above the first row of siding, then cut and fit T1-11 siding for the gable ends. Cover the horizontal seam with 1 × 4 trim snugged up against the flashing.

To complete the trim details, add 1 × 2 along the gable ends and sides of the box bay. Use 1 × 4 on all vertical corners and around the windows, service door, and overhead door. Rip down 1 × 10s for horizontal trim along the bottom of the box bay. Also cover underneath the bay joists with ½" exterior-grade plywood.

Rip-cut 1 × 6 boards to 4⅛" wide for the overhead door jambs. Install the jambs using the door manufacturer's dimensions for the opening. Shim behind the jambs if necessary. Make sure the jambs are flush with the inside of the wall framing and extend ⅝" beyond the outside of the framing. Install the 2 × 6 trim as shown in the OVERHEAD DOOR HEADER DETAIL and OVERHEAD DOOR JAMB DETAIL (page 205).

Install the two windows and the service door following the manufacturers' instructions. Position the jambs of the units so they will be flush with the siding, if applicable. Install the overhead door, then add stop molding along the top and side jambs; see the SERVICE DOOR HEADER/JAMB DETAIL (page 205).

Install ½" plywood roof sheathing, starting at the bottom ends of the rafters. Add building paper and asphalt shingles.

Kit Sheds

The following pages walk you through the steps of building two new sheds from kits. The metal shed measures 8 × 9 feet and comes with every piece in the main building pre-cut and pre-drilled. All you need is a ladder and a few hand tools for assembly. The wood shed is a cedar building with panelized construction—most of the major elements come in preassembled sections. The wall panels have exterior siding installed, and the roof sections are already shingled. For both sheds, the pieces are lightweight and maneuverable, but it helps to have at least two people for fitting everything together.

As with most kits, these sheds do not include foundations as part of the standard package. The metal shed can be built on top of a patio surface or out in the yard, with or without an optional floor. The wood shed comes with a complete wood floor, but the building needs a standard foundation, such as wooden skid, concrete block, or concrete slab foundation. To help keep either type of shed level and to reduce moisture from ground contact, it's a good idea to build it over a bed of compacted gravel. A 4-inch-deep bed that extends about 6 inches beyond the building footprint makes for a stable foundation and helps keep the interior dry throughout the seasons.

Before you purchase a shed kit, check with your local building department to learn about restrictions that affect your project. It's recommended—and often required—that lightweight metal sheds be anchored to the ground. Shed manufacturers offer different anchoring systems, including cables for tethering the shed into soil, and concrete anchors for tying into a concrete slab.

Kit sheds offer the storage you need, a quick build, and an attractive addition to your backyard. The metal shed kit shown here is constructed on pages 218 to 221.

Prefabricated sheds need not be plain. If you have the budget, a modern backyard wonder such as this unit is available ready to assemble. The manufacturer of this shed offers the option of buying it ready-to-assemble, or hiring inside an "installers network" to have a professional build it on site for you.

Building a Metal or Wood Kit Shed

If you need an outbuilding but don't have the time or inclination to build one from scratch, a kit shed is the answer. Today's kit sheds are available in a wide range of materials, sizes, and styles—from snap-together plastic or resin lockers to Norwegian pine cabins with divided-light windows and loads of architectural details. Equally diverse is the range of quality and prices for shed kits. One thing to keep in mind when choosing a shed is that much of what you're paying for is the materials and the ease of installation. Better kits are made with quality, long-lasting materials, and many come largely preassembled. Most of the features discussed below will have an impact on a shed's cost.

The best place to start shopping for shed kits is on the Internet. Large manufacturers and small-shop custom designers alike have websites featuring their products and available options. A quick online search should help you narrow down your choices to sheds that fit your needs and budget. From there, you can visit local dealers or builders to view assembled sheds firsthand. When figuring cost, be sure to factor in all aspects of the project, including the foundation, extra hardware, tools you don't already own, and paint and other finishes not included with your kit.

High-tech plastics, such as polyethylene and vinyl, are often combined with steel and other rigid materials to create tough, weather-resistant—and washable—kit buildings.

If you're looking for something special, higher-end shed kits allow you to break with convention without breaking your budget on a custom-built structure.

Here are some of the key elements to check out before purchasing a kit shed:

Materials

Shed kits are made of wood, metal, resin, various plastic compounds, or any combination thereof. Consider aesthetics, of course, but also durability and appropriateness for your climate. For example, check the snow load rating on the roof if you live in a snowy climate, or inquire about the material's UV resistance if your shed will receive heavy sun exposure. The finish on metal sheds is important for durability. Protective finishes include paint, powder-coating, and vinyl. For wood sheds, consider all of the materials, from the framing to the siding, roofing, and trimwork.

Extra Features

Do you want a shed with windows or a skylight? Some kits come with these features, while others offer them as optional add-ons. For a shed workshop, office, or other workspace where you'll be spending a lot of time, consider the livability and practicality of the interior space, and shop accordingly for special features.

What's included?

Many kits do not include foundations or floors, and floors are commonly available as extras. Other elements that may not be included:

- Paint, stain, etc.—Also, some sheds come pre-painted (or pre-primed), but you won't want to pay extra for a nice paint job if you plan to paint the shed to match your house.

- Roofing—Often the plywood roof sheathing is included but not the building paper, drip edge, or shingles.

Most shed kits include hardware (nails, screws) for assembling the building, but always check this to make sure.

Assembly

Many kit manufacturers have downloadable assembly instructions on their websites, so you can really see what's involved in putting their shed together. Assembly of wood sheds varies considerably among manufacturers—the kit may arrive as a bundle of pre-cut lumber or with screw-together prefabricated panels. Easy-assembly models may have wall siding and roof shingles already installed onto panels.

Extenders

Some kits offer the option of extending the main building with extenders, or expansion kits, making it easy to turn an 8 × 10-ft. shed into a 10 × 12-ft. shed, for example.

Foundation

Check with the manufacturer for recommended foundation types to use under their sheds. The foundations shown in the Building Basics section (page 14) should be appropriate for most kit sheds.

Shed hardware kits make it easy to build a shed from scratch. Using the structural gussets and framing connectors, you avoid tricky rafter cuts and roof assembly. Many hardware kits come with lumber shopping lists so you can build the shed to the desired size without using plans.

How to Build a Metal Kit Shed

Prepare the building site by leveling and grading as needed, and then excavating and adding a 4"-thick layer of compactible gravel. If desired, lay landscape fabric under the gravel to inhibit weed growth. Compact the gravel with a tamper and use a level and a long, straight 2 × 4 to make sure the area is flat and level.

NOTE: Always wear work gloves when handling shed parts—the metal edges can be very sharp. Begin by assembling the floor kit according to the manufacturer's directions—these will vary quite a bit among models, even within the same manufacturer. Be sure that the floor system parts are arranged so the door is located where you wish it to be. Do not fasten the pieces at this stage.

Once you've laid out the floor system parts, check to make sure they're square before you begin fastening them. Measuring the diagonals to see if they're the same is a quick and easy way to check for square.

Fasten the floor system parts together with kit connectors once you've established that the floor is square. Anchor the floor to the site if your kit suggests. Some kits are designed to be anchored after full assembly is completed.

5 Begin installing the wall panels according to the instructions. Most panels are predrilled for fasteners, so the main trick is to make sure the fastener holes align between panels and with the floor.

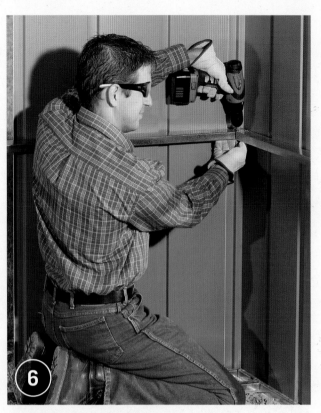

6 Tack together mating corner panels on at least two adjacent corners. If your frame stiffeners require assembly, have them ready to go before you form the corners. With a helper, attach the frame stiffener rails to the corner panels.

7 Install the remaining fasteners at the shed corners once you've established that the corners all are square.

8 Lay out the parts for assembling the roof beams and the upper side frames and confirm that they fit together properly. Then, join the assemblies with the fasteners provided. *(continued)*

Attach the moving and nonmoving parts for the upper door track to the side frames if your shed has sliding doors.

Fasten the shed panels to the top frames, making sure that any fastener holes are aligned and that crimped tabs are snapped together correctly.

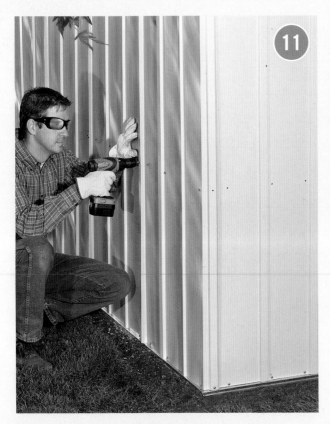

Fill in the wall panels between the completed corners, attaching them to the frames with the provided fasteners. Take care not to overdrive the fasteners.

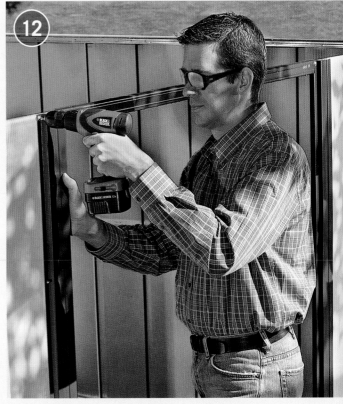

Fasten the doorframe trim pieces to the frames to finish the door opening. If the fasteners are colored to match the trim, make sure you choose the correct ones.

Insert the shed gable panels into the side frames and the door track and slide them together so the fastener holes are aligned. Attach the panels with the provided fasteners.

Fit the main roof beam into the clips or other fittings on the gable panels. Have a helper hold the free end of the beam. Position the beam and secure it to both gable ends before attaching it.

Drive fasteners to affix the roof beam to the gable ends and install any supplementary support hardware for the beam, such as gussets or angle braces.

(continued)

Begin installing the roof panels at one end, fastening them to the roof beam and to the top flanges of the side frames.

Apply weatherstripping tape to the top ends of the roof panels to seal the joints before you attach the overlapping roof panels. If your kit does not include weatherstripping tape, look for adhesive-backed foam tape in the weatherstripping products section of your local building center.

As the overlapping roof panels are installed and sealed, attach the roof cap sections at the roof ridge to cover the panel overlaps. Seal as directed.

NOTE: Completing one section at a time allows you to access subsequent sections from below so you don't risk damaging the roof.

Attach the peak caps to cover the openings at the ends of the roof cap and then install the roof trim pieces at the bottoms of the roof panels, tucking the flanges or tabs into the roof as directed. Install a plywood floor, according to manufacturer instructions.

Assemble the doors, paying close attention to right/left differences on double doors. Attach hinges for swinging doors and rollers for sliding doors.

Install door tracks and door roller hardware on the floor as directed, and then install the doors according to the manufacturer's instructions. Test the action of the doors and make adjustments so the doors roll or swing smoothly and are aligned properly.

TIPS FOR MAINTAINING A METAL SHED

Touch up scratches or any exposed metal as soon as possible to prevent rust. Clean the area with a wire brush, and then apply a paint recommended by the shed's manufacturer.

Inspect your shed once or twice a year and tighten loose screws, bolts, and other hardware. Loose connections lead to premature wear.

Sweep off the roof to remove wet leaves and debris, which can be hard on the finish. Also clear the roof after heavy snowfall to reduce the risk of collapse.

Seal open seams and other potential entry points for water with silicone caulk. Keep the shed's doors closed and latched to prevent damage from wind gusts.

ANCHOR THE SHED

Metal sheds tend to be light in weight and require secure anchoring to the ground, generally with an anchor kit that may be sold separately by your kit manufacturer. There are many ways to accomplish this. The method you choose depends mostly on the type of base you've built on, be it concrete or wood or gravel. On concrete and wood bases, look for corner gusset anchors that are attached directly to the floor frame and then fastened with landscape screws (wood) or masonry anchors driven into concrete. Sheds that have been built on a gravel or dirt base can be anchored with auger-type anchors that are driven into the ground just outside the shed. You'll need to anchor the shed on at least two sides. Once the anchors are driven, cables are strung through the shed along the roof channels and over interior beams (see page 248, for more information on installing anchors).

How to Build a Large Wood Kit Shed

TOOLS & MATERIALS

Phillips screwdriver	Tin snips	3¼ lbs. galvanized nails
Cordless power drill & bits	Chalkline	7 bundles 3-tab roofing shingles
Hammer	Utility knife	1 roll #15 roofing felt
Carpenter's level	Caulk gun	60 ft. drip edge
Carpenter's pencil	Ladder	2 gallons exterior latex paint
Tape measure	Nail gun	1 quart exterior latex trim paint
Framing square	Caulk	Wood glue

A large shed kit such as this makes wonderful use of economies of scale and promises an almost custom-built appearance. When working on a large kit shed such as this, it's essential to use a helper—or more than one.

Level the site and position and level the 4 × 4 treated wood skids or pour concrete footings. Lay the floor frame on a clean, flat surface according to manufacturer's instructions. Nail the ledgers to the joists, and then position the frame on the skids. Center the floor frame flush with the skids at the ends, with the sides overhanging. Check for level both ways, and measure diagonals to ensure square. Attach the floor frame end joists on one end into the skids. Check again for square and level, and attach the opposite end of the frame to the skids.

Nail the floor panels in place, rough-side up, starting on one end and using 3" nails, spaced every 6" around the edges. Snap chalk lines at joists and along the length of the floor every 12 inches. Nail at the intersections. Check the floor for level and square when you're done.

Frame the left wall on the floor. Determine the front-door side and build the wall as you did the floor frame. Nail wall panels onto the framing, finished side out. The panels should be flush to the top and sides of the wall, and overhanging at the bottom. Flip the wall on its face, to the left of the floor frame. Repeat the process with the right wall.

Construct the two "T"s to form the rear wall. Position the shorter T over the longer one, so that the legs are perfectly aligned, with ½" space between the bottom of the top T and the top of the bottom T. (Tack a temporary brace to keep the Ts oriented correctly.) Nail the wall panels on each side of the Ts. Flip the wall, and screw the header to the top plate. Lay the rear wall face down behind the door frame.   *(continued)*

Frame the front door wall according to the manufacturer's instructions. Check the opening for square with a framing square. Nail the front wall panels to the top flush with the top edge to the top plate, and so the inside edges are flush to the studs. (The panels should overhang the bottoms and outside edges.) Center the header on the top plate and attach it to the plate. Tack a temporary brace across the door opening.

Set the rear wall header on the sidewall top plates. Screw the rear wall to the sidewall top plates and nail the bottom edge of the rear wall panels into the floor, using 3" nails spread 6" apart. Toenail the rear wall studs to the floor, and screw the cross brace to the sidewall studs on both sides. Screw the rear wall panel to the sidewall studs. Remove the temporary sidewall braces. Place the front wall in position and screw the top plate into the sidewall top plates. Nail the bottom edge of the front wall panels to the floor frame as you did with the rear wall panels, and nail the front wall to the sidewall end studs.

Tack the temporary jig guide to the floor along the rear wall, 107⅛" from one sidewall sole plate, to create the rafter jig. Position two rafter chords in the jig. Lay a bead of construction adhesive for the gusset, and nail the gusset in place. Flip the rafter and repeat the process on the opposite side. Repeat with the remaining rafters.

Position a rafter over the front sidewall studs, flush to the inside of the front wall. Toe-screw the rafter to the top plates from both sides. Install the remaining rafters in the same way.

9

Assemble the front wall gable on the floor. Assemble the rear wall gable in the same way. Caulk above the wall panels and nail the front gable flush to the top of the wall panels on the bottom and slightly offset along the peak of the rafter. Nail the rear gable Z-strip over the header and rear wall panels. Nail the rear gable flush on the bottom to the top of the wall panels.

10

Screw the loft supports in place between the sidewall top plates. Screw the supplied shelf supports between the studs closest to the rear wall, and toe-screw the crossbrace to the supports. Slide the shelf in the gap between the rear wall top and bottom studs, and resting on the shelf brace. Screw the front edge and nail the rear edge down. Position the loft panel on the loft supports.

11

Attach the rear gable trim pieces with screws from the inside. Screw the front wall overhang trim and pieces onto the front gable.

12

Screw the soffit nailer strips in place, centered under the rafter ends on each side. Toenail the strip to the wall studs. Screw soffit blocks to the rafter ends at each corner. Screw the soffit panels to each block, nailer, and rafter end. Nail the rear wall soffit caps, primed-side out, to the rear wall.

(continued)

Install the first roof panel horizontally and square the roof by nailing the panel to the inside rafter. Lever the panel's outside edge flush with the roof peak and gable's back edge. Nail it down and install the remaining panels and roofing materials.

Finish installing the trim. Install windows and doors and caulk all seams and gaps in preparation for painting.

Paint shed walls, trim, and doors with exterior paint. If you are up for the challenge, try using a paint sprayer for the walls.

Screw weatherstripping in place inside the door frame and along the door edges. Predrill holes and screw the top and bottom door bolts into the door and frame. Screw the decorative hinges horizontally over the cross braces. Drill the hole for the T handle and install the handle.

Turn your attention to the interior. Sheet vinyl or vinyl flooring planks are good choices if you want to install floor coverings, because they are durable and weather-resistant (and pretty easy to install).

Make and hang custom shelving inside the shed—this is a great way to use leftover lumber.

Decorate your shed if you wish to use it as a retreat or workshop. A little paint and a few custom touches will make your new kit shed a favorite getaway for anyone in the family.

How to Build a Wood Kit Shed

 ## PREPARE FOR THE DELIVERY

Panelized shed kits are shipped on pallets. The truck may have a forklift for the driver to unload whole pallets. Otherwise, you'll have to unload the pieces. Use helpers to unload the truck. Once unloaded, carry the pieces to the site and stack them on bolsters. Arrange the stacks according to the assembly steps.

Cut the 4 x 4 or 6 x 6 pressure-treated timbers to form a skid frame to match the outside dimensions of the shed frame. Check the skid frame for level both ways and square, and adjust as necessary.

Prepare the base for the shed's wooden skid foundation with a 4" layer of compacted gravel. Make sure the gravel is flat, smooth, and perfectly level.

NOTE: For a sloping site, a concrete block foundation may be more appropriate (check with your shed's manufacturer and see page 14).

Pile the frame sections next to the skid frame. Begin laying the pieces out on the skid frame, alternating large and small sections to create three full-width sections.

Screw the frame sections together. Check the frame for level and then fasten it to the skid frame following the manufacturer's recommendations.

Screw the plywood to one end of the frame for the floor, starting with a large sheet at the left rear corner. Screw down the two outside deck boards. Lay out the remaining boards using spacers between the boards to maintain even spacing.

Lay out the shed's wall panels in their relative positions around the floor. Make sure they are right-side up: with the windows on the top half. On windowless panels, the siding indicates which end is up.   *(continued)*

Position the two rear corner walls upright onto the floor so the wall is flush with the floor's edges. Screw the wall panels together. Raise and join the remaining wall panels one at a time. Do not fasten the wall panels to the shed floor yet.

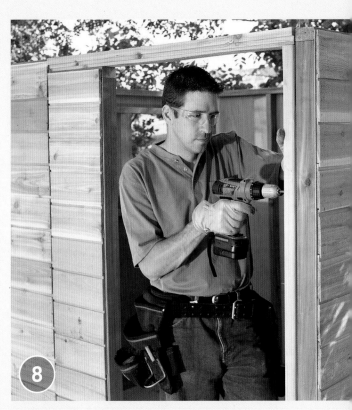

Screw the door header on top of the narrow front wall panel flush with the wall. Fasten the header with screws. Screw the door jamb to the right side wall framing to create a ½" overhang at the end of the wall. Fasten the header to the jamb.

Confirm that all wall panels are properly positioned. The walls should be flush with edges of the floor frame; the wall siding overhangs the floor. Screw the wall panels through the sole plate down into the floor frame.

Install the wall's top plates starting with the rear wall. Install the side-wall plates as directed—these overhang the front of the shed and will become part of the porch framing. Finally, install the front wall top plates.

Assemble the porch rail sections using the screws provided for each piece. Attach the top plate extension to the 4 × 4 porch post, and then attach the wall trim/support to the extension. Fasten the corner brackets, centered on the post and extension. Install the handrail section 4" up from the bottom of the post.

Install each of the porch rail sections. Fasten through the wall trim/support and into the side wall, locating the screws where they will be least visible. Fasten down through the wall top plate at the post and corner bracket locations to hide the ends of the screws. Anchor the post to the decking and floor frame with screws driven through angled pilot holes.

Hang the Dutch door. Install the hinges onto the door panels. Use three pairs of shims to position the bottom door panel: ½" shims at the bottom, ⅜" shims on the left side, and ⅛" shims on the right side. Fasten the hinges to the wall trim/support. Hang the top door panel in the same fashion, using ¼" shims between the door panels.

Join the two pieces to create the rear wall gable, screwing through the uprights on the back side. On the outer side of the gable, slide in a filler shingle until it's even with the neighboring shingles. Fasten the filler with two finish nails located above the shingle exposure line, two courses up. Attach the top filler shingle with two (exposed) galvanized finish nails. *(continued)*

Position the rear gable on top of the rear wall top plates and center it from side to side. Use a square or straightedge to align the angled gable supports with the angled ends of the outer plates. Screw the gable to the plates and wall framing. Assemble and install the middle gable.

Arrange the roof panels on the ground according to their installation. Flip the panels over and screw framing connectors to the rafters at the marked locations.

With one or two helpers, set the first roof panel at the rear of the shed, then set the opposing roof panel in place. Align the ridge boards of the two panels, and then screw them together. Do not fasten the panels to the walls at this stage.

Position one of the middle roof panels, aligning its outer rafter with that of the adjacent rear roof panel. Screw the rafters together. Install the opposing middle panel in the same way. Set the porch roof panels into place one at a time—these rest on a ½" ledge at the front of the shed. From inside the shed, screw the middle and porch panels together along their rafters.

Check the fit of all roof panels at the outside corners of the shed. Make any necessary adjustments. Screw the panels to the shed, starting with the porch roof. Inside the shed, fasten the panels to the gable framing, then anchor the framing connectors to the wall plates.

Install the two roof gussets between the middle rafters of the shed roof panels (not the porch panels): First measure between the side walls—this should equal 91" for this kit (see Resources, page 250). If not, have two helpers push on the walls until the measurement matches your requirement. Hold the gussets level, and screw them to the rafters.

(continued)

Add filler shingles at the roof panel seams. Slide in the bottom shingle and fasten it above the exposure line two courses up, with two screws. Drive the screws into the rafters. Install the remaining filler shingles the same way. Attach the top shingle with two galvanized finish nails.

Cover the underside of the rafter tails (except on the porch) with soffit panels, fastening to the rafters with finish nails. Cover the floor framing with skirting boards, starting at the porch sides. Hold the skirting flush with the decking boards on the porch and with the siding on the walls, and screw them in place.

Add vertical trim boards to cover the wall seams and shed corners. The rear corners get a filler trim piece, followed by a wide trim board on top. Add horizontal trim boards at the front wall and along the top of the door. Fasten all trim with finish nails.

At the rear of the shed, fit the two fascia boards over the ends of the roof battens so they meet at the roof peak. Fasten the fascia with screws. Install the side fascia pieces over the rafter tails with finish nails. The rear fascia overlaps the ends of the side fascia. Cover the fascia joints and the horizontal trim joint at the front wall with decorative plates.

Screw the two roof ridge caps along the roof peak, overlapping the caps' roofing felt in the center. Install the decorative gusset gable underneath the porch roof panels using mounting clips. Finish the gable ends with two fascia pieces installed with screws.

Complete the porch assembly by screwing each front handrail section to a deck post. Fasten the handrail to the corner porch post. The handrail should start 4" above the bottoms of the posts, as with the side handrail sections. Anchor each deck post to the decking and floor frame with screws (see Drilling Counterbored Pilot Holes, below).

 ## DRILLING COUNTERBORED PILOT HOLES

Use a combination piloting/ counterbore bit to pre-drill holes for installing posts. Angle the pilot holes at about 60°, and drive the screws into the framing below whenever possible. The counterbore created by the piloting bit helps hide the screw head. As an alternative, you can use an angled drilling jig.

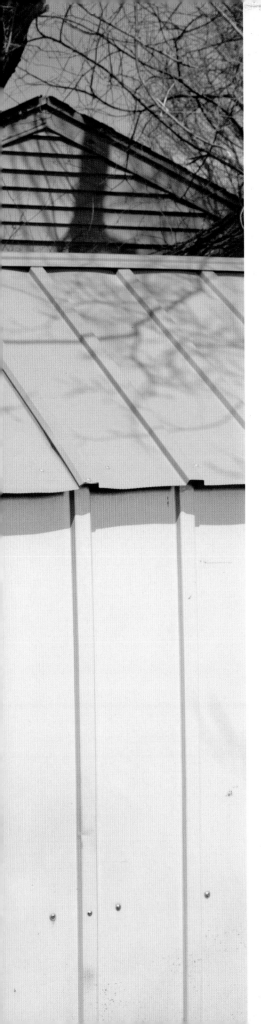

Shed Maintenance & Repair

Maintaining your shed is crucial to getting the most for your money out of this workhorse structure. How well you take care of it will also determine whether you have a focal point or an eyesore in your yard.

Shed maintenance begins with the area around the shed. Regularly check to make sure trees and bushes are not encroaching on the shed or damaging any part of it. This includes roots that may be growing under the foundation. Trim back plants or overhanging branches that threaten the shed.

Keep the shed roof free of leaves and yard debris, and clean dirt and debris out of the door tracks to avoid damaging the tracks or the doors. Regularly lubricate door slides or wheel axles with silicone lubricant.

However, the biggest enemy of wood and metal sheds is water. To avoid rot and rust, keep woodpiles and other structures away from the sides of the shed, and ensure that the area around the foundation allows for efficient drainage. Also keep an eye on the shed: regularly check wood surfaces with a wood probe to detect any rot as soon as it arises, and inspect your metal shed all around at least twice a year.

Even taking all these precautions, you'll probably still have to deal with one condition or another over the course of your shed's life. Whether you need to jack up the sinking foundation, bring new life to old doors or install new, you'll find easy-to-follow instructions in the pages that follow. Make repairs quickly, as soon as any problem arises, and they'll be easy to deal with.

In this chapter

Painting a Metal Shed

The paint on a metal shed is likely to show wear and tear long before the integrity of a metal shed is compromised. Fortunately, fading or distressed paint is a simple thing to fix. Paint your shed regularly to ensure the longest life and the nicest appearance in the yard. Although aluminum can be treated with aluminum siding brightener, the steel that forms the construction of most prefab sheds will need to be sanded and painted.

Always use a quality primer meant for metal. and prepare the surface of the shed well so that the primer and paint go on as smooth as possible. Sand off any rust you find. Although it's easiest for most home DIYers to paint with a brush and roller, if you're comfortable with a paint sprayer, rent one from a local tool rental center—the job will go much more quickly. Just be careful to rent the right filter for the paint—latex or oil-based—that you're using.

TOOLS & MATERIALS

Sandpaper

Cleaner

Metal primer

Exterior metal paint

Roller, roller sleeve and brush, or paint sprayer

Work gloves

Eye protection

New paint is not just a way to preserve the investment you've made in your shed; it's also a way to bring the shed back to life, and create a much more attractive view in the yard.

How to Paint a Metal Shed

1

Clear all debris from the shed. If the shed has been waxed, clean thoroughly with automotive wax remover or aluminum siding cleaning solution.

2

Sand, clean, and prime all rusted areas with a primer rated for exterior metal. If rot extends through any part of the shed's metal, cut the rot out and patch the area.

3

Paint the shed. A roller and brush may be the best tools for a small shed in a confined yard. If you have experience with a paint sprayer, you'll get excellent results in less time—as long as you're spraying on a day with no wind.

Jacking Up a Shed

The sinking shed is a common problem. Unless a shed is built on cement piers that extend below the frost line, it is likely that sooner or later the shed will sink if the ground is subject to freeze/thaw cycles. Usually one side or one corner will sink faster. The easiest solution is to jack up that side or corner of the shed and add some shimming material to level it. Ground contact lumber or cement blocks can be used as shims. You may decide to add another set of support blocks or skids. Empty the shed of its contents to make the project easier.

You can rent a bottle jack for this repair; most rental centers offer them. However, depending on what type of jack your car has, and how heavy your shed is (if it's heavier than your car, don't use the jack), you may be able to use the car jack in the same way the bottle jack is used in the steps that follow. If the skids or joists have rotted, they should be replaced. This is a more involved process and not covered here.

Leveling a sunken shed is a surprisingly simple fix. Making the repair is your chance to improve drainage away from the shed, and is an easy way to increase the longevity of your shed.

Before

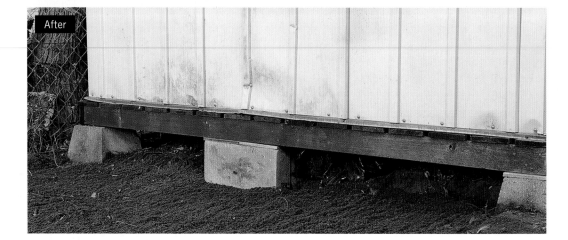
After

Using nails and hammer, install a masonry string level across the base of the shed. Nails should be equidistant from the top and bottom of the baseboards.

Excavate the area around the shed base to expose the foundation, and clear the area to determine the problem; sometimes a sinking shed is due to rotting skids, not settling soil. Remove debris or dirt so you have a clear view of the foundation materials. Depending on the landscape, you may have to dig out an additional area to install the jack.

Dig a trench if you will need room to maneuver a jack handle. The jack needs a firm surface beneath it to dissipate the pressure. Stack two 12 × 12" squares of plywood under the jack. Make sure the wood above the jack is solid. Place a chunk of 2 × 4 or 2 × 6 lumber on top of the jack. Slowly pump up the jack. It is best not to move structures more than ¼" at a time. Allow the structure to rest for a day after each raising.

When the string reads level, insert the appropriate shimming material between the skid or joists and the gravel or concrete base. Ground-contact treated lumber or solid cement bricks or blocks are good choices.

 SAFETY TIP

To avoid crushing and related injuries, never work on, under, or around a load supported only by a jack. Always use jack stands, or solid cement bricks or blocks, to support the shed.

Installing Wooden Doors on a Metal Shed

One of the biggest frustrations of owning a metal shed is that the doors often become unusable long before the shed itself is ready for replacement. The doors are usually the weakest points of these structures, with slides that tend to bind and metal that flexes and leads to the doors regularly jamming.

Whether your metal doors have become unusable or you just want to upgrade the look of your metal shed, adding hinged wood doors can be a major upgrade. The simple doors shown here are easy to construct and simple to install, and will last every bit as long as the shed itself. For simplicity's sake, we've used plywood and left the doors unfinished. But you can always use higher-quality boards or paint the doors for even more design options. Regardless of what look you go for, the doors look and perform much better than the standard metal versions provided with most prefab sheds.

TOOLS & MATERIALS

Cordless drill
2 × 4 lumber
Plywood siding
Deck screws
Sheet metal screws
Angle brackets
Galvanized butt hinges
Work gloves
Eye protection

Steel shed doors often expire long before the shed itself. A new set of wood doors is an opportunity for a stylish upgrade.

How to Install Wooden Doors on a Metal Shed

Sliding doors are typically removed by unfastening the screws attaching the door to the top slides then tilting back and lifting the door out of the bottom track.

Use 2 × 4s to frame a door opening. Measure and cut two boards to the height of the opening. Align them in the shed door opening, and use blocking to attach these at the top to the roof beams. Attach a 2 × 4 between the two sides as a header. Attach the base of the side 2 × 4s to the bottom door track, using angle brackets and sheet metal screws.

Measure the opening, and create two doors of equal size to fit into it. Use 2 × 4s for framing, with a diagonal brace, and cover with plywood siding. Attach the doors to the frame using galvanized butt hinges. Attach hardware.

Replacing Rotted or Damaged Shed Wood

Sometimes even the best maintenance doesn't stop the wood in a shed from rotting or deteriorating. Powerful direct sun on walls facing south can wreak havoc on an otherwise well-maintained shed. Woodpiles, shrubs, or snow piles can all lead to contact rot that often goes unnoticed until it's too late. Rot problems can even be a consequence of design, when wall, window, or door surfaces include flat, horizontal members where moisture can collect.

The best possible approach is to remove and replace all the affected wood, whether you're replacing siding, door wood, windowsills, or any other part of the shed. Keep in mind that most types of wood siding are not rated for ground contact. As part of making wood repairs on sheds, it's wise to ensure the drainage away from the shed is adequate, and make changes as necessary. This may involve the addition of gutters or rain chains in some cases, because splashing from water running off the roof is often the cause of moisture and rot problems.

To keep ahead of rot and moisture problems, finish or treat all surfaces of siding, doors, and other wood shed surfaces. Do this on a yearly basis and you'll be taking good steps to ensure that you don't need to repeat the repairs in the pages that follow.

PARTIAL REPLACEMENT

Alternately, you could remove only the bottom, rotted portion of the siding. This is only a short term fix, as this partial job is susceptible to future problems. Remove trim boards and mark a level, horizontal line about 6" above the rotted area. Use a circular saw set to the thickness of the siding and saw along the line. Remove the rotted wood. Cut a new piece of siding to fit. Insert a piece of Z-flashing under the old siding. Slide the new siding under the flashing and nail to studs. Replace the trim.

Accumulating rot defaces and devalues your shed; fortunately there's a quick fix.

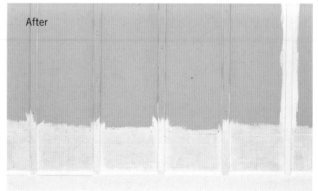

Replacing rotting siding is a common fix for wooden sheds. After painting it will look as good as new.

How to Replace Rotting Siding

1

Remove the trim boards and/or batten with a pry bar and hammer to gain access to the rotting portion. The corner boards may also have to be removed.

2

Determine where the damaged or rotted material ends and snap a chalk line at least 6" past that point to serve as your cutting line. Where possible, snap chalk lines that fall midway across a framing member.

3

Use a circular saw to make a straight cut along the chalk line. The saw setting should be just slightly deeper than the width of the panel (usually about a ½"). It is easiest and safest to make the cuts before removing the fasteners that hold the panels in place. Finish the cut with a jigsaw or handsaw where the circular saw cannot reach. Remove any screws and pry off the damaged material.

4

Cut a 2 × 4 nailer to create a surface to which the replacement wood can be secured. To secure the blocking, use a handheld drill and drive deck screws toenail style into the studs. The backers will be used to secure the replacement patch.

5

Measure and cut replacement panels using the same size and type of material. To allow for expansion and contraction, leave no more than ⅛" gap between the patch board and the original siding. Use screws or nails to fasten the patch. Screws provide better holding power but are more difficult to conceal. Reattach trim and then prime and paint all exposed wood surfaces.

6

TIP: Make corrective repairs to fix problems so they don't recur. Use a pneumatic nailer to secure appropriate molding to the ledge of the base trim. Here, pieces of ¾ × ¾" quarter-round molding are set into thick beds of caulk on the top edges of the base trim and then secured with finish nails. This creates a surface that sheds water instead of allowing it to accumulate.

How to Repair a Wood Shed Door

Locate and mark framing members inside the wall (use a stud finder) so you can draw cutting lines around the damage that falls over studs. Starting at the bottom, cut clapboards at the cutting lines with a keyhole saw or a wallboard saw. For access, slip wood shims under the clapboard above the one you're cutting.

NOTE: The repair will look better if you stagger cutting lines so they don't fall on the same stud.

Cut replacement clapboards to fit, using a miter saw or power miter saw.

Nail the replacement clapboards in the patching area (you can use tape to hold them in place if you like). Follow the same nailing pattern used for the boards around it. Set nail heads with a nailset.

Caulk the gaps between clapboards and fill nailholes with exterior putty or caulk. Prime and paint the repaired section to match.

 ## WOODEN SHED MAINTENANCE

- Trim shrubs and tree branches
- Sweep leaves and twigs off roof
- Inspect roofing for breaks and cracks

- Inspect siding for damage or rot
- Inspect flooring for damage or rot
- Oil door hinges

How to Repair Board & Batten Siding

With a flat pry bar, remove the battens on each side of the damaged area. To protect the painted surfaces, cut along the joints between the boards and battens with a utility knife before removing the battens.

Remove the damaged panel (or make vertical cuts with a circular saw underneath the batten locations if any of the battens are only decorative). Cut a replacement panel from matching material, sized to leave an ⅛" gap between the original panels and the new patch. Nail the panel in place, caulk the repair seams and reinstall the battens. Prime and paint to match.

How to Repair Tongue & Groove Siding

Back of groove area on patch is removed

To repair tongue-and-groove siding, first mark both ends of the damage and rip-cut the board down the middle, from end to end, with a circular saw or trim saw. Then use the trim saw to cut along the vertical lies. Finish all cuts with a keyhole saw. Split the damaged board with a pry bar or a wide chisel. Then pull the pieces apart and out of the hole. If the building paper below the damaged section was cut or torn, repair it.

Cut the replacement board to fit, then cut off the backside of the groove so the board can clear the tongue on the course below. Prime the board, let it dry, then nail it into place. Paint to match.

Installing an Anchor Cable

If you live in an area subject to high winds and your shed is not anchored to a concrete foundation, it should be anchored with a cable anchoring system. Most shed manufacturers sell a variety of anchoring systems for their sheds. The cable system is the easiest to install after construction. If you live in hurricane- or tornado-prone areas, note that local ordinances may cover what sort of shed anchoring system is allowable. For the anchoring system to work properly, the shed needs to be level and firmly supported. If it isn't, fix those issues before installing the cable anchors.

TOOLS & MATERIALS

Anchor kit (see Resources, page 250)

Adjustable wrench

Work gloves

Eye protection

Wire cutters or bolt cutters (optional)

Any shed not tied to a permanent foundation can usually benefit from anchoring. The anchors are a simple, inexpensive way to ensure the shed is not damaged or moved during high winds. Anchors are required by code in hurricane- and tornado-prone parts of the country.

How to Install an Anchor Cable

Align the anchors 5" to 9" from the side of the shed and parallel with the first roof panel. Twist the anchors into the ground using a short rod or crowbar through the eye of the anchor. Twist until 3" of the anchor extends above ground.

Measure and cut the anchor cable as necessary. (If you are retrofitting an older shed or a using anchors from a different manufacturer, the anchors may be supplied with a single long cable that you cut to accommodate your needs. Use wire cutters or bolt cutters to cut the anchor cable.) Insert the cable under the roof panel at the rib, then over all the roof beams and out the opposite side along the rib.

Insert about 6" of cable through the eye of the anchor and attach the cable using the cable clamp. Pull the cable somewhat taut, and repeat with the anchor on the opposite side. Repeat with the second set of anchors. Do not overtighten, as you can damage the shed. Check the cables annually for tautness and loosen or tighten the anchors as needed.

Resources

Arrow Storage Products
1101 North 4th St.
Breese, IL 62230
(800) 851-1085
www.arrowsheds.com

Best Barns
PO Box 995
Carbondale, CO 81623
(800) 987-4337
www.bettersheds.com

Cedarshed Industries
#215 - 20780 Willoughby Town Centre Drive
Langley, BC
Canada V2Y 0M7
(800) 830-8033
www.cedarshed.com

Jamaica Cottage Shop
170 Winhall Station Road
South Londonderry, VT 05155
(802) 297-3760
www.jamaicacottageshop.com

Lifetime Products
P.O. Box 160010
Freeport Center Bldg. D-11
Clearfield, UT 84016
(800) 225-3865
www.lifetime.com

Outdoor Living Today
9393 287th St.
Maple Ridge, BC
Canada V2W1L
(888) 658-1658
www.outdoorlivingtoday.com

Reeds Ferry Small Buildings, Inc.
3 Tracy Lane
Hudson, NH 03051
(603) 883-1362
www.reedsferry.com

Summerwood Products
735 Progress Ave.
Toronto, ON Canada
Canada, M1H 2W7
(866) 519-4634
www.summerwood.com

Sun-In-One
500 Philadelphia Pike
Wilmington, DE 19809
(877) 280-9473
www.suninone.com

Suncast Homeplace Collection
301 Commerce Drive, Suite 400
New Holland, PA 17557
(866) 768-8465
www.homeplacestructures.com

Tuff Shed
1777 S. Harrison St, Suite 600
Denver, CO 80210
(866) 860-8833
www.tuffshed.com

Photo Credits

Alamy
p. 35 (bottom left, Ian Francis), 148 (top, Francisco Martinez)

Clive Nichols
p. 33 (bottom right)

Dency Kane
p. 150 (bottom), 152 (bottom right)

Douglas Keister
p. 81 (bottom)

Dreamstime
p. 82 (bottom right)

iStockPhoto
p. 80 (top)

Malco Tools
p. 46 (bottom)

Outdoor Living Today
p. 32, 33 (top), 34 (top, bottom), 72 (bottom right), 81 (lower right), 82 (top left), 96 (bottom left)

Photolibrary
p. 33 (bottom left, Garden Pix LTD), 82 (bottom right, Jason Smalley), 146 (Per Magnus)

Reeds Ferry Small Buildings, Inc.
p. 79 (bottom)

Rubbermaid
p. 80 (bottom left)

Sheds Unlimited
p. 78

Shutterstock
p. 149 (top, bottom)

Studio Shed
p. 35 (top), 83 (bottom), 99 (top), 213

Summerwood
p. 81 (top right), 83 (top), 147, 150 (top), 151, 152 (top), 153 (top, bottom)

Sun-in-One
p. 138, 139 (top and bottom)

Metric Conversion Charts

CONVERTING MEASUREMENTS

TO CONVERT:	TO:	MULTIPLY BY:
Inches	Millimeters	25.4
Inches	Centimeters	2.54
Feet	Meters	0.305
Yards	Meters	0.914
Square inches	Square centimeters	6.45
Square feet	Square meters	0.093
Square yards	Square meters	0.836
Cubic inches	Cubic centimeters	16.4
Cubic feet	Cubic meters	0.0283
Cubic yards	Cubic meters	0.765
Pounds	Kilograms	0.454

TO CONVERT:	TO:	MULTIPLY BY:
Millimeters	Inches	0.039
Centimeters	Inches	0.394
Meters	Feet	3.28
Meters	Yards	1.09
Square centimeters	Square inches	0.155
Square meters	Square feet	10.8
Square meters	Square yards	1.2
Cubic centimeters	Cubic inches	0.061
Cubic meters	Cubic feet	35.3
Cubic meters	Cubic yards	1.31
Kilograms	Pounds	2.2

CONVERTING TEMPERATURES

Convert degrees Fahrenheit (F) to degrees Celsius (C) by following this simple formula: Subtract 32 from the Fahrenheit temperature reading. Then multiply that number by $5/9$. For example, 77°F - 32 = 45. 45 × $5/9$ = 25°C.

To convert degrees Celsius to degrees Fahrenheit, multiply the Celsius temperature reading by $9/5$. Then, add 32. For example, 25°C × $9/5$ = 45. 45 + 32 = 77°F.

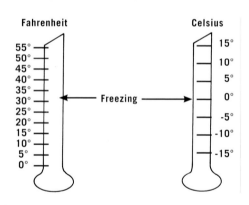

METRIC PLYWOOD PANELS

Metric plywood panels are commonly available in two sizes: 1,200 mm × 2,400 mm and 1,220 mm × 2,400 mm, which is roughly equivalent to a 4 × 8-ft. sheet. Standard and Select sheathing panels come in standard thicknesses, while Sanded grade panels are available in special thicknesses.

STANDARD SHEATHING GRADE		SANDED GRADE	
7.5 mm	($5/16$ in.)	6 mm	($4/17$ in.)
9.5 mm	($3/8$ in.)	8 mm	($5/16$ in.)
12.5 mm	($1/2$ in.)	11 mm	($7/16$ in.)
15.5 mm	($5/8$ in.)	14 mm	($9/16$ in.)
18.5 mm	($3/4$ in.)	17 mm	($2/3$ in.)
20.5 mm	($13/16$ in.)	19 mm	($3/4$ in.)
22.5 mm	($7/8$ in.)	21 mm	($13/16$ in.)
25.5 mm	(1 in.)	24 mm	($15/16$ in.)

LUMBER DIMENSIONS

NOMINAL - U.S.	ACTUAL - U.S. (IN INCHES)	METRIC
1 × 2	¾ × 1½	19 × 38 mm
1 × 3	¾ × 2½	19 × 64 mm
1 × 4	¾ × 3½	19 × 89 mm
1 × 5	¾ × 4½	19 × 114 mm
1 × 6	¾ × 5½	19 × 140 mm
1 × 7	¾ × 6¼	19 × 159 mm
1 × 8	¾ × 7¼	19 × 184 mm
1 × 10	¾ × 9¼	19 × 235 mm
1 × 12	¾ × 11¼	19 × 286 mm
1¼ × 4	1 × 3½	25 × 89 mm
1¼ × 6	1 × 5½	25 × 140 mm
1¼ × 8	1 × 7¼	25 × 184 mm
1¼ × 10	1 × 9¼	25 × 235 mm
1¼ × 12	1 × 11¼	25 × 286 mm
1½ × 4	1¼ × 3½	32 × 89 mm
1½ × 6	1¼ × 5½	32 × 140 mm
1½ × 8	1¼ × 7¼	32 × 184 mm
1½ × 10	1¼ × 9¼	32 × 235 mm
1½ × 12	1¼ × 11¼	32 × 286 mm
2 × 4	1½ × 3½	38 × 89 mm
2 × 6	1½ × 5½	38 × 140 mm
2 × 8	1½ × 7¼	38 × 184 mm
2 × 10	1½ × 9¼	38 × 235 mm
2 × 12	1½ × 11¼	38 × 286 mm
3 × 6	2½ × 5½	64 × 140 mm
4 × 4	3½ × 3½	89 × 89 mm
4 × 6	3½ × 5½	89 × 140 mm

LIQUID MEASUREMENT EQUIVALENTS

1 Pint	= 16 Fluid Ounces	= 2 Cups
1 Quart	= 32 Fluid Ounces	= 2 Pints
1 Gallon	= 128 Fluid Ounces	= 4 Quarts

COUNTERBORE, SHANK & PILOT HOLE DIAMETERS

SCREW SIZE	COUNTERBORE DIAMETER FOR SCREW HEAD (IN INCHES)	CLEARANCE HOLE FOR SCREW SHANK (IN INCHES)	PILOT HOLE DIAMETER	
			HARD WOOD (IN INCHES)	SOFT WOOD (IN INCHES)
#1	.146 (9/64)	5/64	3/64	1/32
#2	1/4	3/32	3/64	1/32
#3	1/4	7/64	1/16	3/64
#4	1/4	1/8	1/16	3/64
#5	1/4	1/8	5/64	1/16
#6	5/16	9/64	3/32	5/64
#7	5/16	5/32	3/32	5/64
#8	3/8	11/64	1/8	3/32
#9	3/8	11/64	1/8	3/32
#10	3/8	3/16	1/8	7/64
#11	1/2	3/16	5/32	9/64
#12	1/2	7/32	9/64	1/8

NAILS

Nail lengths are identified by numbers from 4 to 60 followed by the letter "d," which stands for "penny." For general framing and repair work, use common or box nails. Common nails are best suited to framing work where strength is important. Box nails are smaller in diameter than common nails, which makes them easier to drive and less likely to split wood. Use box nails for light work and thin materials. Most common and box nails have a cement or vinyl coating that improves their holding power.

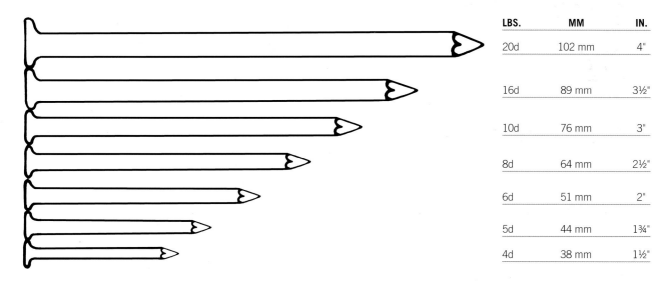

LBS.	MM	IN.
20d	102 mm	4"
16d	89 mm	3½"
10d	76 mm	3"
8d	64 mm	2½"
6d	51 mm	2"
5d	44 mm	1¾"
4d	38 mm	1½"

Index